Practical Radio Engineering and Telemetry for Industry

Titles in the series

Practical Cleanrooms: Technologies and Facilities (David Conway)

Practical Data Acquisition for Instrumentation and Control Systems (John Park, Steve Mackay)

Practical Data Communications for Instrumentation and Control (Steve Mackay, Edwin Wright, John Park)

Practical Digital Signal Processing for Engineers and Technicians (Edmund Lai)

Practical Electrical Network Automation and Communication Systems (Cobus Strauss)

Practical Embedded Controllers (John Park)

Practical Fiber Optics (David Bailey, Edwin Wright)

Practical Industrial Data Networks: Design, Installation and Troubleshooting (Steve Mackay, Edwin Wright, John Park, Deon Reynders)

Practical Industrial Safety, Risk Assessment and Shutdown Systems for Instrumentation and Control (Dave Macdonald)

Practical Modern SCADA Protocols: DNP3, 60870.5 and Related Systems (Gordon Clarke, Deon Reynders)

Practical Radio Engineering and Telemetry for Industry (David Bailey)

Practical SCADA for Industry (David Bailey, Edwin Wright)

Practical TCP/IP and Ethernet Networking (Deon Reynders, Edwin Wright)

Practical Variable Speed Drives and Power Electronics (Malcolm Barnes)

Practical Radio Engineering and Telemetry for Industry

David Bailey BE (Comms) BAILEY AND ASSOCIATES, PERTH, AUSTRALIA

Newnes

OXFORD AMSTERDAM BOSTON HEIDELBERG LONDON NEW YORK
PARIS SAN DIEGO SAN FRANCISCO SINGAPORE SYDNEY TOKYO

Newnes
An imprint of Elsevier
Linacre House, Jordan Hill, Oxford OX2 8DP
200 Wheeler Road, Burlington, MA 01803

First published 2003

British Library Cataloguing in Publication Data
A catalogue record for this book is available from the British Library

ISBN   07506 58037

For information on all Newnes publications, visit
our website at www.newnespress.com

Typeset and Edited by Vivek Mehra, Mumbai, India

Printed and bound in Great Britain

Preface

This book covers the fundamentals of telemetry and radio communications, describes their application and equips you with the skills to analyze, specify and debug telemetry and radio communications systems.

The structure of the book is as follows.

Chapter 1: Radio technology. This chapter goes through the fundamentals of radio theory and then introduces all the elements of a complete radio system. Finally a coherent methodology will be provided to systematically design, install, and test a successful radio system for use in telemetry systems.

Chapter 2: Line of sight microwave systems. In this chapter the theory of radio link transmission and the components which make up a microwave system are described.

Chapter 3: Satellite systems. This chapter will examine the types of satellite services that are available, the different service providers and their satellites, the satellite frequency bandplan, the fundamentals of satellite systems, their operation and then, finally, look at the various satellite services available along with their relevance to telemetry.

Chapter 4: Reliability and availability. This chapter looks at reliability with relevance to an overall telemetry installation and then will look at availability, with respect to the various types of telemetry communication links that have been discussed.

Chapter 5: Infrastructure requirements for master sites and RTUs. This chapter identifies the different issues such as location selection, lightning protection, equipment shelters, power supplies and voice and data cabling involved in setting up a master radio telemetry site.

Chapter 6: Integrating telemetry systems into existing radio systems. This chapter describes what is involved in implementing telemetry systems into existing radio systems ranging from high to low integrity categories.

Chapter 7: Miscellaneous telemetry systems. Three different telemetry systems which differ from standard telemetry systems are discussed here.

Chapter 8: Practical system examples. This chapter reviews two typical applications of radio communications for a dockside operation and a remote oceanographic sensor system.

Contents

1

Radio technology

1.1 Introduction

A significant number of telemetry systems use radio as a communications medium. It is often chosen as a communications medium in preference to using cables (landlines) for a number of reasons:

- Telemetry systems are often operated over large distances, and the costs of installing cables can far exceed the costs of installing radio equipment
- If landlines are to be rented from a telephone company, it is often found that the costs of buying radio equipment are amortized in several years while rental costs continue (this can depend on whether the lines are switched or dedicated)
- Depending on the type of equipment being installed and the distance involved, radio can generally be installed faster than other communication mediums
- Radio equipment is very portable and can be easily moved to new locations as when plants are relocated. If a plant or equipment goes out of service, the radio equipment can be reused in other locations with little or no modification
- The user can own the radio communication links, which allows him to transmit any information in any format that he requires (assuming that all physical and regulatory constraints are taken into consideration)
- Reasonably high data rates are available
- The radio unit can be used as a backup to landline circuits where high integrity communication circuits are required

Radio can be broken down into two general classifications:

- Equipment that operates below 1 GHz, and
- Equipment that operates above 1 GHz

The latter is generally referred to as microwave radio and will be discussed in detail, in Chapter 2. The former will be referred to as just **radio** in this book. This chapter is devoted to developing an understanding, of the concepts of radio and how these can be practically applied to the implementation of a telemetry system.

The science of radio communications, and how it can be successfully implemented into a practical solution, in an industrial environment, is often considered a difficult and esoteric task that is best left to the experts. Many readers of this book would know of a radio system installation that does not work well or suffers from intermittent problems. This should not be the case if a system is engineered correctly! A logical methodology can be used in designing radio links for telemetry.

This chapter will first go through the fundamentals of radio theory and then introduce all the elements of a complete radio system. Finally, a coherent methodology will be provided to systematically design, install, and test a successful radio system for use in telemetry systems.

1.2   Fundamentals of radio operation

A single unshielded conductor that carries an alternating current (ac) and has a certain amount of internal resistance will radiate energy in the form of **electromagnetic waves**. The frequency of the radiated waves is equal to the oscillation rate of the alternating current. For example, an alternating current that has electrons moving backwards and forwards along a conductor at a speed of 100 oscillations per second will radiate electromagnetic waves at a frequency of 100 hertz (Hz).

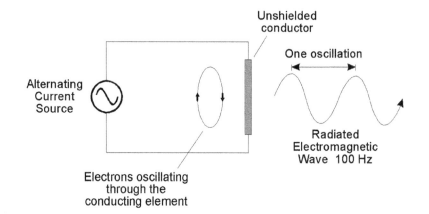

Figure 1.1
Production of an electromagnetic wave

Hertz is the term used to describe the oscillation of electrons, in oscillations per second, or the number of complete cycles of an electromagnetic wave, in cycles per second. Cycles/second and hertz are used interchangeably.

The following abbreviations are used to describe frequency:

- kHz   kilohertz   = 1×10^3 Hz
- MHz   Megahertz   = 1×10^6 Hz
- GHz   Gigahertz   = 1×10^9 Hz

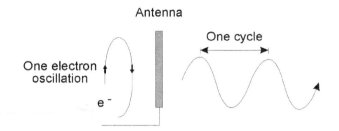

Figure 1.2
One cycle

One cycle is one wavelength. The effectiveness, of the element as a radiator (antenna) of electromagnetic waves, is dependant upon its degree of resonance at the oscillating frequency. (Resonance occurs when the energy injected into a load is absorbed by the load because the particles that the load is constructed of oscillate at the same frequency as the injected energy.)

Electromagnetic waves are purportedly very small particles called photons (much smaller than an electron) that travel in a close sinusoidal pattern through space. There are two branches of complex physics that study the nature of electromagnetism, **particle physics,** and **wave mechanics**. Each theory studies the phenomenon from a different perspective, sometimes with quite different results. This book will avoid any involvement in the long convoluted mathematics of these studies, adhering to the more practical realities of working with radio.

Electromagnetic waves can be represented by an electric field and a magnetic field. The electric field is produced by the moving electrons in the conductor. The electric field itself appears to produce an equivalent magnetic field perpendicular to the electric field. Figure 1.3 illustrates the components of an electromagnetic wave.

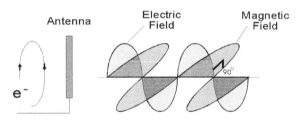

Figure 1.3
Construction of an electromagnetic wave

With the antenna of a radio, the electric field of an electromagnetic wave is parallel to the antenna elements; that is, it is parallel to the moving electrons. The relevance of this is discussed further in section 1.13.

When electromagnetic waves are traveling through space and pass through a conducting element, if the electrical field is parallel to the conductor, the wave will cause electrons in the conductor to move up and down in sympathy with the incoming sinusoidal wave. This conductor would be the receiving antenna, which feeds the retrieved signals through to the receiver. The receiver would detect the movement of electrons and process it as the incoming signal.

The mediums of air and vacuum are quite often referred to as **free space**.

Electromagnetic waves travel at the speed of light (3×10^8 meters/sec) through a vacuum. The more dense the medium through which the waves are required to travel, the more they slow down. The speed of electromagnetic waves in air is almost the same as through a vacuum.

For determining the wavelength of a certain frequency the following formula is used:

$$C = \lambda f$$

Where:
  C = speed of electromagnetic waves in meters per second
  λ = wavelength of signal in meters
  f = frequency of signal in hertz.

For example, if we have a radio system operating at 900 MHz then the wavelength of this signal is 33 cm.

Electromagnetic waves that have a short wavelength (high frequency) tend to travel in a straight line and are quickly absorbed or reflected by solids. Long wavelength (low frequency) electromagnetic waves tend to be more affected by atmospheric conditions and travel a more curved path (that follows the curvature of the earth). They are also more able to penetrate solids. In this chapter, short wavelengths will be regarded as frequencies from 335 MHz up to approximately 960 MHz and long wavelengths being frequencies of 1 MHz up to approximately 225 MHz.

Electromagnetic waves, in the radio frequency and microwave frequency bands (10 kHz to 60 GHz), are often just referred to as RF signals (radio frequency signals).

1.2.1 Components of a radio link

A radio link consists of the following components:

- Antennas
- Transmitters
- Receivers
- Antenna support structures
- Cabling
- Interface equipment

Figure 1.4 illustrates how these elements are connected together to form a complete radio link.

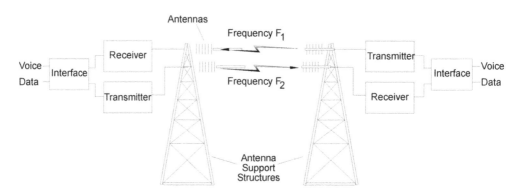

Figure 1.4
Fundamental elements of a radio link

Antenna

This is the device used to radiate or detect the electromagnetic waves. There are many different designs of antennas available. Each one radiates the signal (electromagnetic waves) in a different manner. The type of antenna used depends on the application and on the area of coverage required.

Transmitter

This is the device that converts the voice or data signal into a modified (modulated) higher frequency signal and feeds it to the antenna, where it is radiated into the free space as an electromagnetic wave, at radio frequencies.

Receiver

This is the device that converts the radio frequency signals (fed to it from the antenna detecting the electromagnetic waves from free space) back into voice or data signals.

Antenna support structure

An antenna support structure is used to mount antennas in order to provide a height advantage, which generally provides increased transmission distance and coverage. It may vary in construction from a three-meter wooden pole to a thousand-meter steel structure.

A structure, which has guy wires to support it, is generally referred to as a **mast**. A structure, which is freestanding, is generally referred to as a **tower**.

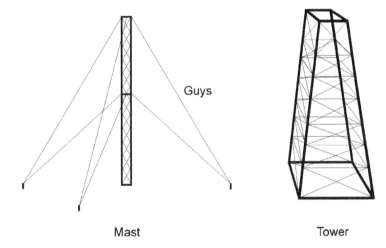

Figure 1.5
Illustration of mast and tower

Cabling

There are three main types of cabling used in connecting radio systems:

- Coaxial cable for all radio frequency connections
- Twisted pair cables for voice, data, and supervisory connections
- Power cables

Interface equipment

This allows connection of voice and data into transmitters and receivers, from external sources. It also controls the flow of information, timing of operation on the system, and control and monitoring of the transmitter and receiver.

1.3 The radio spectrum and frequency allocation

1.3.1 General

There are very strict regulations that govern the use of various parts of the radio frequency spectrum. Specific sections of the radio frequency spectrum have been allocated for public use. All frequencies are allocated to users by a government regulatory body. Figure 1.6 illustrates the typical sections of the radio spectrum allocated for public use around the world. Each section is referred to as a **band**.

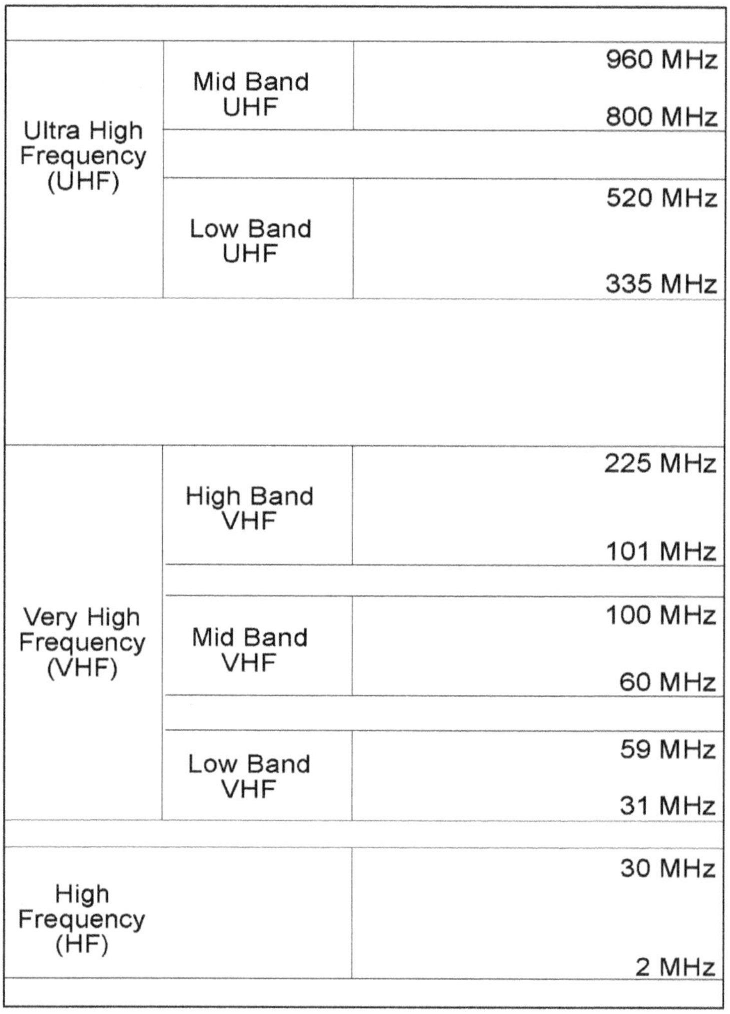

Ultra High Frequency (UHF)	Mid Band UHF	960 MHz 800 MHz
	Low Band UHF	520 MHz 335 MHz
Very High Frequency (VHF)	High Band VHF	225 MHz 101 MHz
	Mid Band VHF	100 MHz 60 MHz
	Low Band VHF	59 MHz 31 MHz
High Frequency (HF)		30 MHz 2 MHz

Figure 1.6
The radio spectrum for public use

Certain sections of these bands will have been allocated specifically for telemetry systems.

In some countries, a deregulated telecommunications environment has allowed sections of the spectrum to be sold to large private organizations that manage it and in turn, sell to (smaller) individual users.

An application is made to a government body, or independent groups that hold larger chunks of the spectrum for reselling, to obtain a frequency license as no transmission is allowed on any frequency unless a license is obtained.

1.3.2 Single and two frequency systems

The governing body will allocate one or two frequencies, per channel, to the user. In the radio bands, each frequency will generally be 12½ or 25 kHz wide.

In radio terms, this represents the difference between operating a **simplex** or **duplex** system. Note that simplex and duplex have slightly different meanings in radio than in data communication.

1.3.2.1 Single frequency allocation

If a user is allocated a single frequency, then the system is said to be a **simplex** system. There are two modes of simplex radio transmissions:

Single direction simplex

Here, information in the form of single frequency radio waves will travel in one direction only, from a transmitter to a receiver.

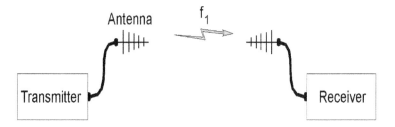

Figure 1.7
Single direction simplex

Two direction simplex

Here, the single frequency is used to transmit information in two directions but only in one direction at a time.

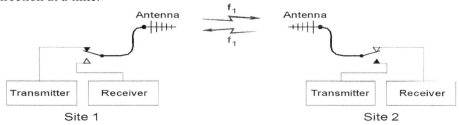

Figure 1.8
Two direction simplex

1.3.2.2 Two frequency allocation

Here two frequencies are allocated, approximately 5 MHz apart, depending on the operating band. This type of system is referred to as a **duplex** system. There are two modes of duplex radio transmission:

Half duplex

Here, one frequency is used for transmission in one direction and the other frequency is for transmission in the other direction, but transmission is only in one direction at one time.

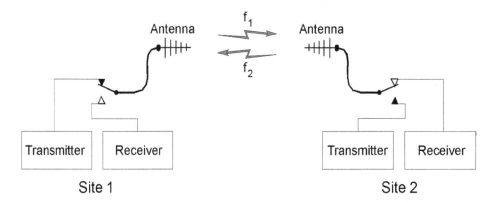

Figure 1.9
Half duplex

This mode is most commonly used for mobile vehicle radio systems, where a radio station is used to repeat transmissions from one mobile to all other mobiles on the frequencies (i.e. talk through repeaters).

Full duplex

Here, one frequency is used for transmission in one direction and the other frequency is for transmission in the other direction, with both transmissions occurring simultaneously.

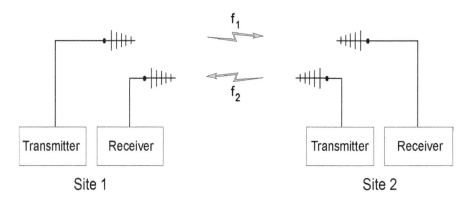

Figure 1.10
Full duplex

1.4 Gain, level, attenuation and propagation

To fully understand the nature of radio communications, it is important to have an understanding of the fundamental parameters used to describe and measure the performance of a radio system. This section will provide details of important measurements and performance parameters associated with radio communication.

1.4.1 Gain and loss

As an electronic signal passes through a circuit or system, the strength of that signal (referred to as its level), will vary. The strength of the signal is measured as the voltage, or current levels of the electrons, at a particular point in the circuit or system. Figure 1.11 below illustrates a device where levels can be measured at three points (A, B and C).

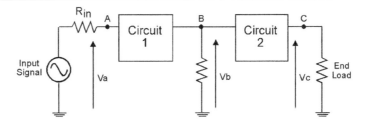

Figure 1.11
Circuit gain and loss

In this case, measurements are made between points A, B or C and earth, and are made in volts, millivolts, or microvolts.

If the voltage level at B is greater than at A then Circuit 1 has provided **GAIN** to the signal. Gain is a measurement of the level at the output of a device compared to the level at the input at the device.
Therefore:

$$GAIN_1 = \frac{V_B}{V_A}$$

If the voltage at B was 10 volts and the voltage at A was 5 volts, circuit 1 would be providing a GAIN factor of **2**. If the voltage at C is less than the voltage at B, then circuit 2 has introduced a LOSS. For example, if the voltage at C is measured as 5 volts, then circuit 2 has introduced a loss factor of **2**

$$LOSS_2 = \frac{V_B}{V_C} = \frac{10}{5} = 2$$

The accepted convention is to always express levels as GAIN and not loss.
Therefore, in the example the gain factor would be ½.

$$GAIN_2 = \frac{V_C}{V_B} = \frac{5}{10} = \frac{1}{2}$$

In radio systems, the measure of the strength of an electromagnetic signal, received by an antenna is expressed as a voltage level, measured within the receiver.

For the radio system, shown in Figure 1.12, the same principles apply. The gain measurement can be applied between any two points in a complete communications system.

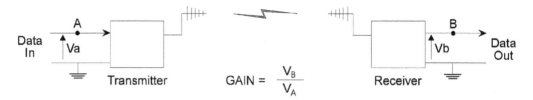

Figure 1.12
System gain for a radio link

1.4.2 Level

The majority of engineering measurements performed on radio systems are carried out as a measurement of power levels. The equations for power are:

$$P = VI$$

$$P = \frac{V^2}{R}$$

$$P = I^2R$$

Where:

P = power (in watts)
V = voltage (in volts)
I = current (in amperes)
R = load resistance

The measurement of level with respect to power originated when Alexander Graham Bell invented a unit of measure for sound levels. This unit became known as the '**bel**'. One tenth of a **bel** is called a **decibel**.

The human ear hears sound in a logarithmic manner. Therefore, a level of 100 watts to the human ear would sound twice as loud as a level of 10 watts (not 10 times).

A one-decibel increase in sound is approximately the smallest increase in sound level, detectable by the human ear.

This unit of measure is now used as the basis for measuring relative power levels in radio, voice, and data networks. For the radio network in Figure 1.12 the gain of the system becomes:

$$\text{GAIN} = Log_{10}\left(\frac{P_\text{B}}{P_\text{A}}\right) \text{bels}$$

$$= 10 \ Log_{10}\left(\frac{P_\text{B}}{P_\text{A}}\right) \text{decibels}$$

Note that this is a relative measurement. The resulting value is a measure of the power level at point B with reference to the power level at point A (power at B relative to power at A). The resultant is NOT an absolute value.

For example, if for the system shown in Figure 1.12 there is an input signal at point A of 1 watt and an output signal at point B of 10 watts, the system gain is:
Written as <u>10 dB</u>.

$$\text{GAIN} = 10 \, Log_{10} \left(\frac{10}{1} \right)$$

$$= 10 \times (1)$$

$$= 10 \text{ decibels}$$

When working with radio equipment, measurements can be made at single points in a system with reference to 1 watt or 1 milliwatt (instead of with reference to a level at another point). The equation then becomes:

$$\text{LEVEL} = 10 \, Log_{10} \left[\frac{P}{1} \right] : \text{dBM}$$

(*With reference to 1 watt*)

or

$$\text{LEVEL} = 10 \, Log_{10} \left[\frac{P}{10^{-3}} \right] : \text{dBM}$$

(*With reference to 1 milliwatt*)

If measurements are required to be carried out in, volts or amperes then replacing power with:

$$\frac{V^2}{R} \quad \text{or} \quad I^2 R$$

$$\text{GAIN} = 10 Log_{10} \left(\frac{V_B^2 . R_A}{V_A^2 . R_B} \right) \text{dB}$$

When $R_A = R_B$

$$\text{GAIN} = 20 Log_{10} \left(\frac{V_B}{V_A} \right) \text{dB}$$

$$\text{GAIN} = 10 \, Log_{10} \left(\frac{I_B^2 R_B}{I_A^2 R_A} \right) \text{dB}$$

or if $R_A = R_B$

$$\text{GAIN} = 20 \, Log_{10} \left(\frac{I_B}{I_A} \right) \text{dB}$$

For most radio systems, R_A will equal R_B and the second formula can normally be used.

Voltages are also sometimes given in decibel forms, where they are measured with respect to 1 volt or 1 microvolt i.e. dBV or dBµV respectively.

1.4.3   Attenuation

The term attenuation is used to express loss in a circuit. Negative gain is therefore considered to be attenuation.

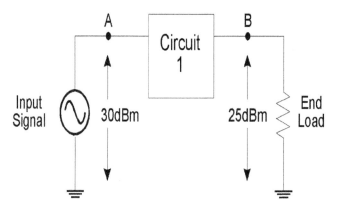

Figure 1.13
Attenuation

With reference to Figure 1.13, the output signal is 5 dBm less than the input signal, and therefore the system has an attenuation of 5 dB (NOT 5 dBm). Attenuation, like gain, is simply a relative measure of output level compared to input level.

1.4.4   Propagation

The methods and parameters involved in transferring electromagnetic waves from one point to another point some distance away, and the way in which these electromagnetic waves are affected in their traveled path by the environment, are embraced by the study of electromagnetic wave propagation. Because there are so many environmental factors that influence the propagation of electromagnetic waves, a high degree of uncertainty exists in determining the reliability of a signal. When designing a radio system, the engineering and prediction of a radio path performance, is quite often the most difficult and involved aspect of radio design.

There are a number of modes of propagation of radio waves across the planet. The mode that takes place and the losses associated with the mode are affected by the following factors:

- Frequency used
- Terrain type
- Time of year
- Weather conditions
- Moisture and salt content of the terrain
- Distance of propagation
- Antenna heights
- Antenna types and polarization

The various modes of propagation are described below:

1.4.4.1 Surface wave

Here waves travel across the surface of the earth to the destination. The wave appears to hug the earth's surface as it moves across it and will provide communication below the horizon.

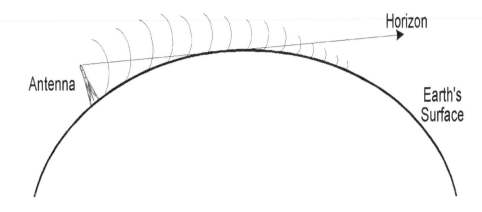

Figure 1.14
Surface waves

1.4.4.2 Ionospheric reflection and scatter

Here radio waves reflect off the ionosphere (a layer of the atmosphere where the air molecules have been ionized by the sun and free electrons exist) back to the earth's surface. Generally, multiple reflections of a single radio wave take place causing, what is referred to as, **SCATTER**. Here communication well below the horizon is possible.

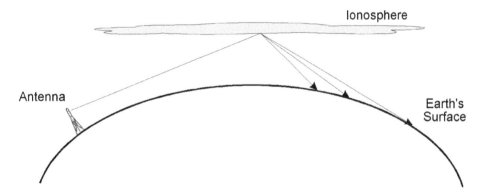

Figure 1.15
Ionospheric reflection and scatter

1.4.4.3 Ionospheric refraction

Here, due to the sudden changes in the characteristics of the atmosphere where the lower atmosphere meets the ionosphere, the radio waves will bend in an arc back to the earth's surface. Again, communication below the horizon is possible.

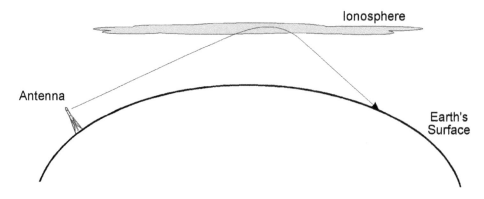

Figure 1.16
Ionospheric refraction

1.4.4.4 Tropospheric scatter

Here, radio waves reflect off a layer of the atmosphere called the TROPOSPHERE, back to earth. The troposphere is the layer of the atmosphere where temperature decreases with height, and this is where most clouds are normally formed. All *weather,* as a general term, originates in the troposphere. Because of the many particles in the troposphere, multiple reflections of a single radio wave occur. This is tropospheric scatter.

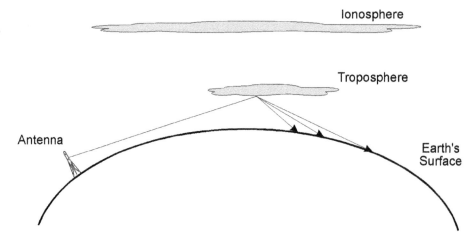

Figure 1.17
Tropospheric scatter

1.4.4.5 Line of sight

Here radio waves travel in an approximate straight line to the destination. There is slight bending of radio waves due to refraction of the earth's atmosphere. Although this is only slight, it does allow the radio wave to travel a little over the straight line to the horizon. The previous four modes of propagation were able to travel significantly over the straight line to the horizon. Light rays seen by the human eye will behave in a similar manner to RF waves. Hence this mode of propagation is referred to as 'line of sight' propagation.

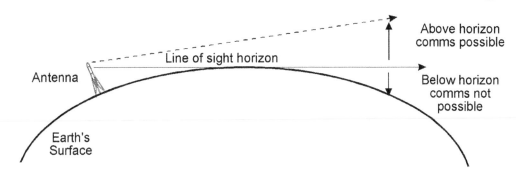

Figure 1.18
Line of sight

1.4.4.6   Diffraction

Radio waves, in a similar manner to light, are able to bend around corners or obstructions. This phenomenon is referred to as diffraction. The degree of diffraction depends on the frequency used and the terrain over which the radio wave is passing. This allows below horizon communications for a line of sight system but introduces significant signal attenuation.

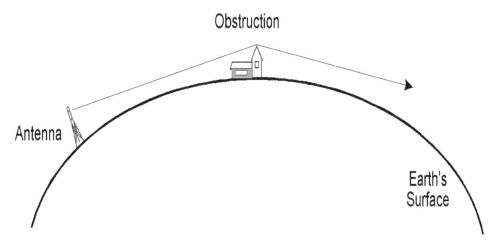

Figure 1.19
Diffraction

1.4.4.7   Ducting

In some parts of the world where large arid landmasses meet the ocean, large areas of temperature inversion can occur in the lower part of the troposphere. This sets up a duct through which radio waves will reflect and refract, for many hundreds of kilometers.

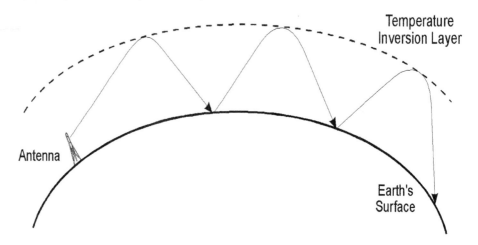

Figure 1.20
Ducting

The majority of telemetry systems are designed to operate in the line of sight mode, and to a lesser extent, in the diffraction and surface wave mode. But the discussion of other propagation modes, as stated above, provide the reader with an understanding as to how interference can be caused by distant users operating on the same frequencies.

1.4.5   Line of sight propagation path attenuation

Under conditions of free space propagation between two antennas that are in line of sight and where the signal is completely unaffected by other environmental factors, the attenuation of the radio wave can be calculated by the following formula:

$$A = 32.5 + 20Log_{10}F + 20Log_{10}D$$
Where:

A = Attenuation in dB
F = Frequency in MHz
D = Distance in km

There are many complex empirical mathematical models that have been developed to describe the different modes of propagation, (discussed in the last section), over different terrains.

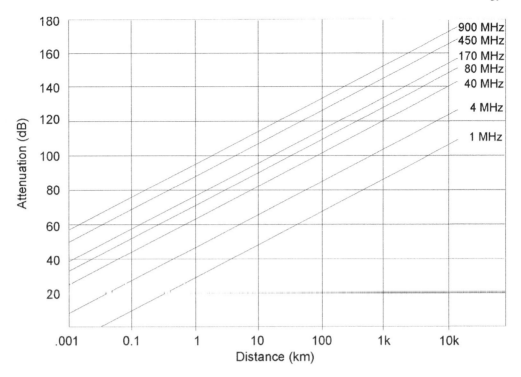

Figure 1.21
Graph of free space attenuation for commonly used radio frequencies

1.4.6   **Reflections**

When setting up telemetry links in the radio frequency band, one of the main causes of degradation of the RF signal will be reflection from the surrounding environment.

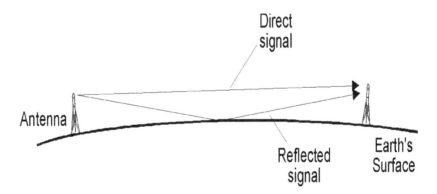

Figure 1.22
Reflected signals

The source of multiple reflections may be the earth, hills, billboards, cars, buildings, airplanes, lakes, rivers (water masses), etc. Reflections from different sources would arrive at the receiving antenna at different times. Because of the different distances each wave may have to travel, they may arrive in or out of phase with the direct signal, which will cause a degree of addition or cancellation of the direct signal respectively. The degree of cancellation will depend on the strength of the reflected signal, which is dependant on the type of surface that reflected it, and the difference in path lengths between the reflected and direct signals, i.e. the phase difference.

1.5   Criteria for selection of frequency bands

Each band of the frequency spectrum behaves differently when used in different physical environments, and therefore, different bands are chosen for specific applications. The choice of correct frequency, for a particular application, is essential in guaranteeing effective and reliable operation of a telemetry system.

The behavioral characteristics of each band will now be considered.

1.5.1   HF

HF radio waves travel around the surface of the earth in two modes. The first is the surface wave or ground wave propagation, as discussed in section 1.4.4. Here the wave front hugs the ground as it follows a curved path across the surface of the earth. The distance, to which the wave can travel and still be at a useable level, depends on the type of terrain and on the conductivity of the earth's surface. The more electrically conductive the surface, the further the ground wave will travel across that surface. For example, the best ground wave propagation occurs over the sea.

Most commercial HF ground wave systems are designed to operate in the frequency band of 1.8 to 3.5 MHz.

Depending on the equipment configuration, distances of approximately 250–500 km are possible over sea, for low power (50–100 watts) transmitters, and approximately 700–1000 km for high power (up to 1000 watts) transmitters. Over relatively flat land, distances of approximately 100–150 km are possible for low power transmitters and approximately 200–300 km for high power transmitters.

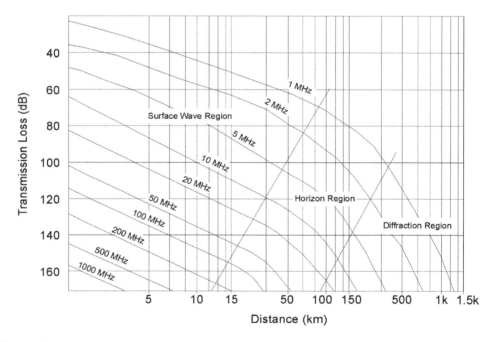

Figure 1.23
Ground wave attenuation for average rolling terrain

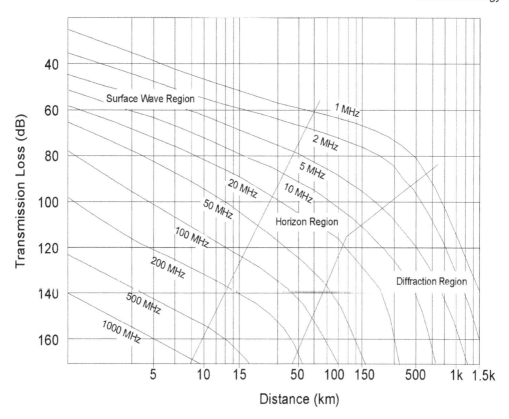

Figure 1.24
Ground wave attenuation for salt water terrain

One important use of the ground wave phenomenon is for communicating with submarines. Here frequencies of 10–30 kHz are used to communicate over sea, spanning distances of many thousands of kilometers. (At these frequencies, there is certain penetration of the sea.)

The second mode of HF propagation is referred to as sky wave communication. This is the Ionospheric mode of propagation, referred to in section 1.4.4. The ionospheric layers are where the electromagnetic waves are affected by the masses of free electrons floating around, because of the ionization of air molecules in these outer layers. There are several layers to the Ionosphere. The first is the E layer, which is approximately 110 km above the earth's surface. Then there are two F layers referred to as F1 and F2 at distances of approximately 230 km and 320 km, respectively. These distances vary depending on the time of day and the two F layers combine at night to form one layer, at approximately 280 km.

An HF radio wave travels from the transmitter to the ionosphere where it then reflects and/or refracts back to the earth's surface. Depending on the type of the earth's terrain, the signal may reflect back up to the ionosphere again. This process may continue until the wave has traveled half way around the earth.

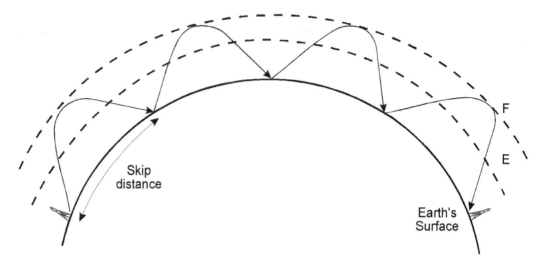

Figure 1.25
Sky wave propagation

The distance, which is covered by a single hop, is referred to as the skip distance. Most HF communication occurs via reflections off the F layers. The maximum distance from a single hop off the F layer is approximately 4000 km. Electromagnetic waves, between the frequencies of 1.8 and 3.5 MHz, are not reflected by the ionosphere and will travel through into space. Therefore, most sky wave communication takes place in the frequency bands above 4 MHz (4–30). The critical frequency of 3.5 MHz is referred to as the absorption limiting frequency (ALF).

For this reason also, ground wave communication is normally carried out between 1.8 and 3.5 MHz. This ensures that a ground wave and a sky wave, from the same transmitter, do not arrive at the same point and cause phase cancellation.

Because of the changing nature of the atmosphere, HF radio transmission is not a very reliable medium of communication. If HF radio is to be used for telemetry purposes then normally only ground wave communication is used, since it provides a signal of higher continuous availability than a sky wave.

The HF radio band is very noisy and because of this, only very slow baud rates can be used (normally 300 or 600 baud).

There is a unique science to predicting and using sky wave communications. Sky wave prediction charts are available each month, showing which frequencies will work and at what time of day these frequencies will operate successfully.

1.5.2   Very high frequency (VHF)

The VHF band covers frequencies from approximately 30 MHz through to 225 MHz. This band is then broken down into three sub-bands as indicated in section 1.3. Each of these sub-bands has slightly different behavioral characteristics.

1.5.2.1   Low band VHF (31—59 MHz)

This band is not commonly used for telemetry system applications. It is considered a high noise band and in particular, is very susceptible to man-made switching and engine-induced noise (though less susceptible than the HF band).

The radio equipment available for use in this band is relatively bulky compared to the equipment used in higher bands. Users generally prefer to set up telemetry systems in higher VHF bands. The higher noise environment of the low band region, limits data speeds (relative to other bands), and raises the minimum receive signal level requirement.

Because these are not popular bands of operation, very few radio manufacturers produce equipment that operates in this band. Being tied to one or two manufacturers will obviously limit the ability to obtain competitive prices for equipment.

There are, however, some advantages in using low band VHF. The lower the VHF frequency used, the more capable the signal is of penetrating solids. For this reason, low band VHF systems are often used in heavily forested areas.

Another advantage is that the lower the VHF frequency used, the better the surface wave propagation, and this is also beneficial in heavily forested areas (although not as effective as HF frequencies). Using line of sight propagation modes, at higher frequencies, for telemetry systems in heavily forested areas has proven to be very unreliable (from the author's own experience). Therefore, the use of low band frequencies is a common solution. In addition, the wet-conditions often encountered in forest areas, provide good soil conductivity and, at low band VHF frequencies, increase the effective antenna height (by providing a good ground plane), and decrease the attenuation of the surface wave. These conditions are significantly less effective for higher VHF bands.

The decision to use low band VHF frequencies in a telemetry system must therefore be carefully considered on the basis of environmental conditions, equipment availability, required data rates and acceptable noise levels.

1.5.2.2 Mid band VHF (60—100 MHz)

Mid band VHF frequencies, are more extensively used than low band VHF frequencies. These frequencies are in a lower noise environment than low band VHF frequencies, and are less affected by switching and engine noise. Therefore, to a limited extent, higher data rates with improved signal availability are possible. Note though, that the noise environment is not as good as that at higher VHF or UHF frequencies.

Although there is considerably more equipment available from manufacturers in this band, it is still limited when compared to the higher frequency bands. Care must be taken when planning a telemetry system in this band to ensure that an adequate range of equipment is available from different manufacturers and that it is locally supported and maintained.

Mid band VHF frequencies, are also relatively good at penetrating solids and are used in some moderately dense forest environments. Wet soil conditions provide no noticeable benefit when operating in this or higher bands. At these higher frequencies, changes in effective antenna height and surface wave attenuation, become negligible.

Because these lower frequencies have good diffraction properties, this frequency band is often used where remote terminal units need to be accessed over hilly terrain, covering large distances. This effect is often utilized in bore-field applications, where it is often not justified to have more than one master site, and some RTUs are very far or partially shadowed by hills. At these, and lower frequencies, the signal will refract around obstructions and to a limited extent, provide a viable communications link. Using the diffraction propagation mode for telemetry system design, calls for extreme caution.

Another noticeable phenomenon, in this frequency band, occurs in parts of the world where large flat dry regions of land meet the ocean. The signal tends to experience long-term fading, lasting several hours or more, because there is a significant drop in received signal strength. This is caused by long-distance ducting due to an atmospheric temperature inversion over the land when the cool air off the ocean hits the hot air on the land and rises above it. Here the direct line of sight or diffracted signal is cancelled by a ducted signal, which will have traveled a greater distance than the direct signal and will arrive out of phase with the direct signal, causing a degree of cancellation.

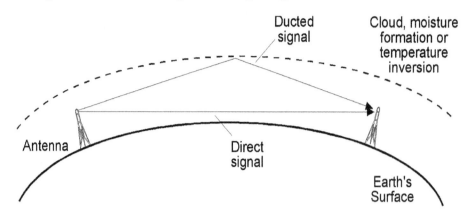

Figure 1.26
Cancellation of signal due to ducting

It is virtually impossible to solve this problem other than to move to a higher frequency band, where ducting does not cause as significant a problem or establish a diverse system.

1.5.2.3 High band VHF (101—225 MHz)

High band is generally the preferred VHF band for telemetry operations. It has the lowest natural noise characteristics and is the least susceptible to externally produced man-made noise. It is generally more reliable when carrying higher data rates because the overall signal availability will be higher than that of the other VHF bands.

Most manufacturers of radio equipment produce equipment that operates in VHF high band. Therefore, mobile radio system implementation, upgrade or replacement, is straightforward and more cost effective than with low or mid VHF bands.

Generally, high band VHF is popular because it provides a reasonable blend of the benefits of lower VHF bands and those of UHF bands.

It will have reasonably good penetration ability through solids and reasonably good diffraction characteristics, but neither the penetration nor the diffraction parameters will be as noticeable as with the lower VHF bands.

In addition, the lower noise communication environment will allow good data transmission rates, but not as good as those available in the UHF band. In built-up urban areas, or on noisy industrial sites, the signal degradation caused by man-made noise can still be quite severe.

The VHF high band generally does not suffer from the major fading problems described for mid band VHF. Minor short term fading, for periods of several seconds, can occur in some locations, but are not as noticeable.

1.5.3 UHF

The UHF band covers the frequencies from approximately 335 MHz through to 960 MHz (the lower parts of the microwave band are also referred to as UHF – *see* Chapter 2). The whole band is generally broken down into two sub-bands having slightly different behavioral characteristics.

1.5.3.1   UHF low band 335—520 MHz

UHF low band frequencies are the ones most commonly used in line of sight telemetry systems. This is because degradation due to noise is less severe than that in the VHF band. The man-made noise from switching equipment and engine ignition, most common in urban areas and on industrial sites, has little effect on UHF frequencies. Therefore, generally higher data rates and lower receive signal levels are possible. A combination of lower possible receiver signal levels and line of sight communications provide a basis for improved radio link availability.

UHF frequencies have minimal penetration ability and, depending on the type of surface the waves are hitting, tend to either be partially absorbed or reflected off the surface. This phenomenon affects telemetry communication systems in two ways. Firstly, in an area such as a city or industrial site where there are a lot of buildings and objects for the signals to reflect off, multiple signals from one transmission may arrive at the receiver. This is referred to as multipathing.

If the transmitter to receiver path is not line-of-sight, perhaps shadowed by a building, there is still some chance that a number of signals will be reflected to the receiver and it will be able to lock onto the strongest reflected signal. Although multipathing can cause cancellation of a direct signal, it is sometimes found to be very short and random in nature and can enhance the communications link. Therefore, uninterrupted communications can sometimes be carried out successfully where there is a lot of random multipathing.

In a more open environment, for example through hilly terrain, there would be fewer but more prominent reflection paths. A reflection off a rock or a pond of water may arrive out of phase and cause severe cancellation of the signal.

When designing a telemetry system operating in the UHF band, the effects of multipathing must be carefully taken into consideration. It is difficult to predict the exact effects of multipathing.

In areas where there is heavy vegetation, the UHF signal tends to be considerably absorbed. It is noted that the wavelength of a UHF signal is close to that of a leaf or branch and when the tree is wet, attenuation becomes even more severe.

UHF frequencies have certain diffraction characteristics but these are significantly less than at the VHF frequencies. Therefore, attenuation of the diffracted signal is significantly increased at the UHF frequencies.

System designers will find that most manufacturers of radio equipment produce a good range of equipment for the UHF band. The author's experience with working in UHF low band is that it is generally very cost effective.

At some locations around the world, on days when the weather is warm and still, significant temperature inversions can occur very close to the surface of the earth and ducting can be experienced over distances of 50 to 150 km. This sometimes causes interference from users operating on the same frequencies at distant locations. This phenomenon is not a major consideration and can be partially overcome using coding techniques in receivers (*See* section 1.21.5).

1.5.3.2   UHF mid band (800—960 MHz)

Frequencies in this band behave in a manner very similar to those in the lower UHF band. In summary when compared to low band UHF they have:

- Higher free space attenuation

- Slightly less signal degradation due to noise and therefore able to carry high data rates at a better availability
- Less penetration ability
- Slightly more reflection ability
- More absorption in vegetation areas – suffer from very high attenuation in dense wet vegetation
- Less diffraction characteristics and higher diffraction attenuation
- More susceptible to reflection cancellation

Equipment manufactured for this band is not as common as for the lower UHF band, but is sufficient and diverse enough to warrant using this band, if required.

1.5.4 Frequency selection

The choice of frequency band that should be used when establishing a telemetry system will depend upon a careful evaluation of a number of criteria. These criteria include:

- Distance to remote sites
- Terrain type
- Vegetation type
- Climate and weather patterns
- Noise environment
- Availability of frequencies
- Availability of equipment
- Required data rates
- Costs

Each of these criteria will now be discussed in more detail.

1.5.4.1 Distance to remote sites

This has been discussed under the sections of propagation, diffraction, surface waves, and the characteristics of different bands. It was seen that for longer distances, lower frequencies should be used and for shorter distances higher frequencies be used. As a rough guide, it could be concluded that for:

- Distances greater than 60 km   Use HF
- Distances between 60 and 30 km   Use VHF
- Distances up to 35 km   Use UHF

Greater distances can be obtained in the VHF and UHF frequency bands depending on the equipment configuration and required link availability (refer to section 4.3).

1.5.4.2 Terrain type

This is a relatively complex consideration. Rarely is the transmission path just a smooth surface, except perhaps over coastal or inland water or over very flat countryside. If the terrain is rolling hills, it is best to use a lower frequency. If the region is mountainous it is often more appropriate to use high UHF frequencies to make use of multipathing effects. Care must be taken to ensure that severe shadowing of an RTU does not occur in a location where no reflected signal can reach.

If the terrain is halfway between mountainous and smooth, rolling hills (referred to as rough terrain), the final determination as to what frequency to use will depend on the earth type and vegetation. If the land is arid, dry, and rocky, it is better to use UHF frequencies. If the land has moderate to heavy vegetation, it is better to use VHF frequencies.

Another factor to consider is the location of stretches of flat land, or water, between the master station and RTUs. These can be sources of significant reflected signals that can cause severe phase cancellations at the receiver.

For example, a radio link operating over a salt lake may operate perfectly well until it rains, at which point the lake will turn into a perfect mirror. The reflection off the lake may cancel the direct signal, to the point that the link drops out.

1.5.4.3 Vegetation type

As was discussed earlier in this section, the denser the vegetation, the lower the frequency that should be used. For example, if transmission is directly through several kilometers of thick, wet forest, it may exhibit an attenuation of 2 dB for a frequency of 30 MHz. For the same section of forest the signal attenuation at 900 MHz may be 40 dB or more.

For thin dry vegetation, the attenuation is noticeable but significantly less than through wet forest, e.g. 5–10 dB at 900 MHz.

1.5.4.4 Climate and weather patterns

The major weather condition that affects the propagation of radio waves in the UHF and VHF bands is the degree of moisture on vegetation; the more rain that falls, the higher the attenuation through vegetation. However, wet surfaces on buildings or on hard rocky mountains, increase the reflective properties of the surfaces and increase the multipath effects, which in most cases would improve the probability of successful communication.

In hot dry regions, significant temperature inversions can occur in the lower parts of the atmosphere during still weather conditions, allowing significant ducting of radio signals, which may cause interference problems at distant receivers.

The presence of fog generally infers a temperature inversion that can cause the same problems.

1.5.4.5 Noise environment

Noise that affects the performance of telemetry systems operating in the VHF and UHF bands is primarily man-made electrical noise. Atmospheric noise only starts to degrade the radio receiver as frequencies go down into low band VHF and HF bands.

Man-made electrical noise can come from switching equipment, relays, rectifiers, inverters, ignition systems, generators, high power ac lines, and numerous other sources. Lightning and atmospheric static build up, produce the worst degree of environmental noise. Static noise is a major problem in the tropical areas of the world.

All noise sources cause degradation of the radio signal at the receiver. They can cause interference with the signal to the point where errors occur in the data. Noise also has the effect of increasing the minimum useable receive signal level, which decreases the availability of the radio path. Therefore, the overall performance of the system degrades.

The degrading effect of man-made noise is worse at lower frequencies. For example, engine noise is quite severe in the 1–225 MHz bands, significantly less severe in the 335–520 MHz band and has virtually no effect in the 800–960 MHz band.

The level of noise is very dependant on the environment. For example, the level of noise in an industrial environment at 450 MHz would be approximately equal to the level of noise in a quiet rural environment at 80 MHz.

Solar noise (also referred to as galactic noise) is noise from space. This is generally considered to be of a low level and only affects, to a small extent, frequencies below 80 MHz. This is worse during the day, when the sun is radiating direct noise, than during the night.

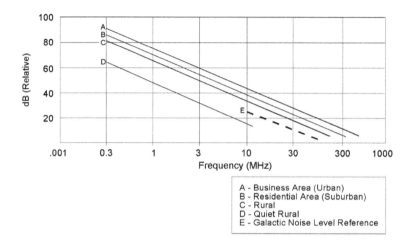

Figure 1.27
Mean values of man-made noise power for different environments (Reference – CCIR Doc 6/167 E/F/s)

Figure 1.27 illustrates relative power levels of man-made noise for different areas (the noise level being relative to thermal noise at 15°C).

Another source of noise occurs when a number of radios operate, within close proximity, on different frequencies. A summation of these frequencies and their associated harmonics form frequencies that interfere with existing frequencies. This is referred to as intermodulation interference and will be discussed in detail in section 1.15.2. It is important that all potential intermodulation noise is determined during the design stage, so that frequencies can be appropriately selected to avoid interference problems.

1.5.4.6 Availability of frequencies

This subject is discussed in detail in section 1.19 under the heading 'Regulatory licensing requirements for radio frequencies'.

1.5.4.7 Availability of equipment

As was discussed in the previous section, equipment operating in lower UHF, high VHF, and HF bands is readily available from different manufacturers, while in mid VHF and high UHF it is sometimes a little harder to obtain and in low VHF band, there is a definite lack of good available equipment on a competitive basis.

During the initial stages of system design, it is essential that the designer determines the availability and range of equipment that can be easily purchased and is fully supported and maintained in his region.

1.5.4.8   Costs

The costs of radio equipment in the 335–520 MHz, 60–100 MHz, and 101–225 MHz bands are generally very competitive. Equipment in the 31–59 MHz and 800–960 MHz band slightly more expensive and equipment in the 1–30 MHz band generally more expensive again.

1.5.5   Summary

The following tables summarize the information that was discussed in this section.

	Low band VHF	**Mid band VHF**	**High band VHF**
Propagation mode	Mostly L.O.S. some surface wave	L.O.S. minimal surface wave	L.O.S.
Data rates	1200 baud	2400 baud	4800 baud
Diffraction properties	Excellent	Very good	Good
Natural noise environment	High	Medium	Low
Affected by man-made noise	Severe	Bad	Some
Penetration of solids	Excellent	Very good	Good
Fading by ducting	Long term	Medium term	Short term
Absorption by wet vegetation	Negligible	Low	Some
Equipment availability	Minimal	Reasonable	Excellent
Relative equipment cost	High	Medium	Low
Uses	– In forested areas – Mostly mobile – Very hilly	– Very hilly & forested areas – Mostly mobile – Over water	– Long distance / L.O.S./hilly areas – L.O.S links – Mobile – Borefields – Over water

Table 1.1

	UHF 1	UHF 2
Propagation mode	L.O.S.	L.O.S.
Data rates	9600 baud	19 200 baud
Diffraction properties	Some	Minimal
Natural noise environment	Low	Negligible
Affected by man made noise	Low	Very low
Penetration of solids	Low	Negligible
Reflection & absorption by solids	Good (enhanced multipathing)	Excellent (excellent multipathing)
Absorption by wet vegetation	High	Very high
Interference by ducting	Some	Some
Equipment availability	Excellent	Reasonable
Relative equipment costs	Low	Medium
Uses	–Telemetry – Mobile	– Telemetry – Mobile – Links

Table 1.2

1.6   Modulation and demodulation

The frequency, at which a radio system operates, is referred to as the carrier wave frequency. If the system has been allocated a frequency of 452.725 MHz, then this is the carrier wave frequency. All information that is to be transferred from the transmitter to the receiver is imparted on to the carrier wave.

Modulation is the process of varying some characteristic of the carrier wave, in accordance with the information signal to be transferred.

Demodulation is the process of deriving the information signal back from the modulated carrier wave.

A modulator is a device that takes the information signal and modulates it on to the carrier. A demodulator, conversely, takes the modulated carrier signal and extracts the information signal out again.

There are four main variations of the modulation techniques used in radio. The first three involve varying the amplitude, frequency, or phase of the carrier in accordance with the information signal. The fourth method is to turn the carrier wave on and off in a digital manner. In any radio system, only one modulation technique is normally used (with a few exceptions). Each of these techniques will now be discussed in detail.

1.6.1   Carrier wave modulation

Carrier wave modulation is where the carrier wave is switched ON and OFF in a digital format. This is used in the simple Morse code telegraphy system.

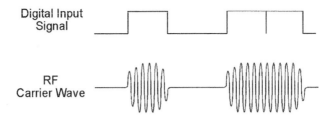

Figure 1.28
Carrier wave modulation

This type of modulation is also referred to as continuous wave (or CW), or ON–OFF keying.

Because of the slow rise and fall times of the output stages of the transmitter, the data speeds possible are severely limited. Generally, this technique is only used for very slow data rates of 50 or 100 baud.

1.6.2   Sidebands and bandwidth

All methods of modulation of a carrier wave produce frequencies that are above and below the carrier frequency. These frequencies are called the sideband frequencies.

Bandwidth is the term used to describe the maximum distance the sideband frequencies are allowed from the carrier. For example, if the allowed bandwidth for an amplitude modulated radio system, working at 1 MHz, is 10 kHz, then the maximum distance the sidebands can extend either side of the carrier is 5 kHz. This is illustrated in Figure 1.29. In this example, the maximum frequency component allowed in the information signal is 5 kHz.

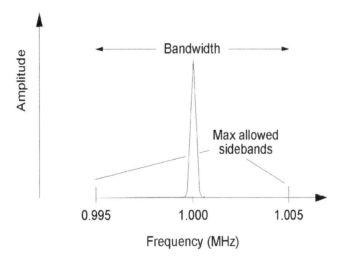

Figure 1.29
Illustration of bandwidth

Normally when describing an operating bandwidth (i.e. that at which a communication system is required to operate on), the outer frequency limits of this bandwidth are the points, where the sideband frequencies have dropped to a maximum power level 3 dB below the maximum central frequency power level.

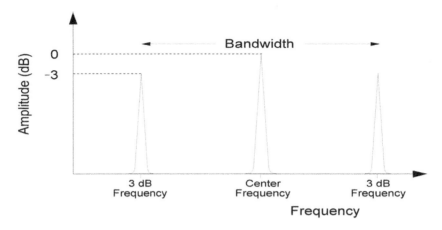

Figure 1.30
3 dB bandwidth

The regulatory government bodies and private organization that allocate frequencies and bandwidth to radio users generally define bandwidth from a different perspective. As an example, they may allocate a new frequency to a user every 25 kHz. An example performance specification would typically specify that for a 25 kHz bandwidth, the output frequency spectrum sideband levels of a transmitter should:

a) For +3 kHz to –3 kHz either side of the carrier have a relatively flat amplitude response

b) For +3 kHz to +6 kHz and –3 kHz to –6 kHz either side of the carrier, the amplitude shall not exceed the levels between +3 kHz and –3 kHz

c) At +6 kHz and –6 kHz either side of the carrier, the amplitude shall be a minimum of 6 dB down on the amplitude of the sidebands at +1 kHz and –1 kHz

d) At frequencies beyond +6 kHz and –6 kHz, the amplitude shall fall off at a rate of 14 dB per octave

e) At ±12.5 kHz, the levels are normally 50 to 70 dB below the ±1 kHz sidebands

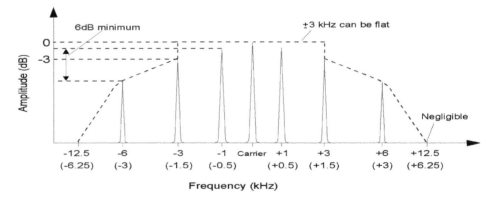

Figure 1.31
Frequency response for 25 kHz and (12½ kHz) bandwidth radio

Although there is still a significant amount of 25 kHz bandwidth allocations around the world, most countries are moving over to 12.5 kHz bandplans, due to a severe shortage of radio spectrum that exists worldwide. In this case, the allocated bandwidth would be the same envelope requirements as for the 25 kHz bandwidth, except that the frequencies either side of the carrier, are halved.

Most manufacturers of radio equipment construct radios so that the transmitter spurious signals are 50–70 dB down outside the allocated bandwidth.

1.6.3   Amplitude modulation (AM)

Amplitude modulation (AM) is the process of varying the amplitude of the carrier wave (which is in sinusoidal form) in sympathy with the information signal. The rate of change (speed) at which the carrier moves up and down in amplitude is directly proportional to the frequency of the information signal. The level of magnitude to which the carrier moves up and down is directly proportional to the amplitude of the information signal.

Figure 1.32 illustrates how a carrier signal is amplitude modulated with an information signal and what the resultant waveform would look like.

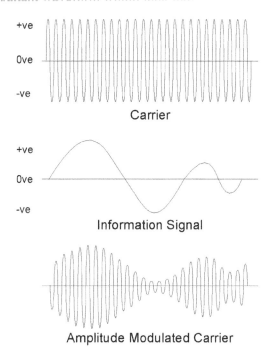

Figure 1.32
The process of amplitude modulation

For amplitude modulation, there will be a sideband frequency, each side of the carrier frequency, for each frequency component in the information signal. For example, if the information signal consists of the frequencies 1 kHz and 2 kHz and it is modulating a carrier wave of 1 MHz, then it will produce sidebands of 0.999 MHz and 1.001 MHz for the 1 kHz signal and 0.998 MHz and 1.002 MHz for the 2 kHz signal. Figure 1.33 illustrates this effect with frequency versus amplitude graphs.

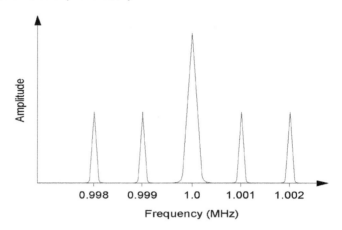

Figure 1.33
AM sidebands produced with a modulating signal that has 1 & 2 kHz frequency components

Note that the amplitude of the carrier frequency component does not change, just the amplitude of the sideband frequency components.

With AM, the difference between the carrier frequency and the farthest sideband frequency component, is determined by the highest frequency component in the information signal.

A common term used in AM is modulation factor, which is a figure used to express the degree of modulation. It is expressed as a percentage of modulation. Figure 1.34 illustrates a number of different modulation factors.

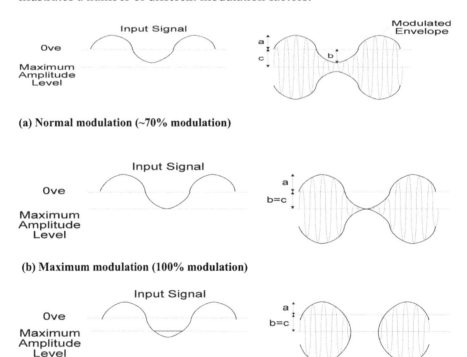

Figure 1.34
Different modulation levels

Figure (a) illustrates what would be an average level input signal modulating a carrier. The percentage modulation is measured as:

$$\% M = \frac{a}{c} \times 100$$

$$\text{OR} \quad \frac{b}{c} \times 100$$

Where *a* and *b* are normally equal.

C is normally referred to as the depth of modulation and is half the maximum modulation.

Figure (b) shows the maximum input signal allowed, which is where $a = b = c$ or where there is 100% modulation. Beyond this modulation-level, the RF wave becomes distorted and will produce spurious sidebands at frequencies beyond the allocated bandwidth. Figure (c) illustrates how over modulation appears at the RF signal.

Spurious sideband frequencies can cause severe interference to nearby receivers, and strict government regulations restrict the user from emitting this interference. A significant proportion of the cost of implementing a radio system can be devoted to filtering any possible sideband interference.

With straight AM modulation a single frequency input signal produces one sideband either side of the carrier. For this reason it is referred to as double sideband amplitude modulation (DSB-AM).

As the carrier and the two sidebands are used to transmit a single piece of information, this is very wasteful of bandwidth resources, considering that all the information is actually contained in a single sideband. Two methods are used to improve the efficiency of the AM system.

With the first method, the carrier is suppressed and only the two sidebands transmitted. This is referred to as double sideband suppressed carrier amplitude modulation (DSBSC-AM). The major advantage of this system is the reduced power requirement at the transmitter output to amplify and transmit the RF signal. In general, this represents a 66% reduction in power requirements. (In a normal DSB-AM circuit, each sideband is a maximum of 25% of the power of the carrier.)

The second method used to increase the efficiency of the radio is to remove the carrier and one of the sidebands. Since both sidebands are carrying the same information, the removal of one sideband does not affect the integrity of the information. This method of modulation is referred to as single sideband suppressed carrier amplitude modulation (SSBSD-AM).

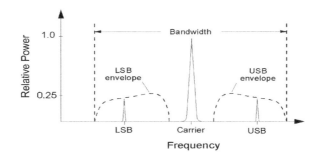

Figure 1.35 – (a) Double sideband amplitude modulation (DSB-AM)

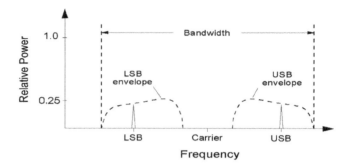

(b) Double sideband suppressed carrier amplitude modulation (DSBSC-AM)

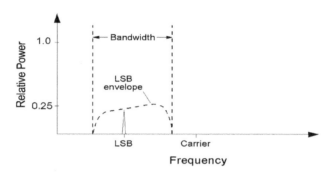

(c) Single sideband suppressed carrier amplitude (SSBSC-AM), upper sideband suppressed

Figure 1.35
Amplitude modulation techniques

Figure 1.35 illustrates the three techniques on amplitude frequency graphs. The graphs illustrate the frequency range over which each sideband is contained. These are referred to as the lower sideband (LSB) envelope and the upper sideband (USB) envelope. Note that for SSBSC systems it is possible to suppress either the USB or the LSB.

Removing one of the sidebands requires only half of the original bandwidth to transmit the same information. This is illustrated in Figure 1.35(c).

AM communications are generally only used on HF systems, with the exception of some important applications, such as aircraft communications (VHF high bands being used). Telemetry systems that use AM mode are only for long distance or surface wave HF communications.

1.6.4   Frequency modulation (FM)

Frequency modulation is the process of varying the *frequency* of the carrier wave, in sympathy with the input signal.

With FM, the frequency of the carrier is continuously changing in sequence with the change in amplitude of the input signal. For example, a low amplitude would correspond to a small change in carrier frequency and a high amplitude to a large change in carrier frequency.

The frequency content of the input signal is conveyed in the RF carrier, as the rate of change of frequency of the carrier. This is illustrated in Figure 1.36.

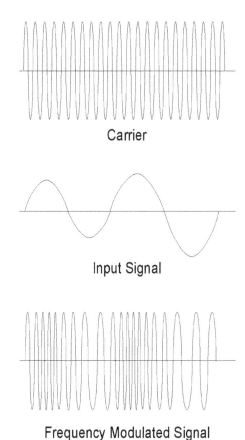

Figure 1.36
Frequency modulation

The amount of frequency the carrier moves off its central frequency is referred to as its **deviation**. Therefore, with reference to Figure 1.37, the deviation is 3 kHz or sometimes expressed as ±3 kHz, where the carrier moves from 1 MHz to 1.003 MHz back to 1 MHz, to 0.997 MHz, back to 1 MHz, up to 1.003 MHz, etc. If the amplitude of the input signal drops by 0.002 MHz, then there will be an approximate drop in deviation to ±2 kHz.

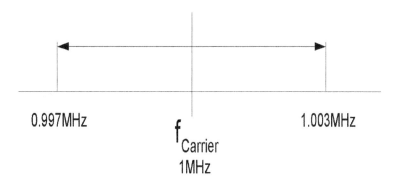

Figure 1.37
Representation of 3 kHz deviation around a 1 MHz carrier

An analogy to help understand frequency modulation is to consider the windscreen wiper of a vehicle. The distance the wiper moves, to either side of the center of the windscreen, is the deviation (i.e. directly proportional to the amplitude of the input signal). The rate of change at which the windscreen wiper travels across the screen (as the wiper will speed up in the middle and then slow down as it reaches the end of each oscillation point), is directly proportional to the frequency of the input signal.

As discussed in section 1.6.2 when a carrier is amplitude modulated with a 1 kHz signal, two sidebands are produced, 1 kHz either side of the carrier. With frequency modulation, an infinite number of sidebands are produced, each being 1 kHz apart on either side of the carrier.

The size of the sidebands varies considerably and depends upon a measure of the modulation referred to as the modulation index. This is expressed as follows:

$$Modulation\ index = \frac{Deviation\ from\ carrier}{Frequency\ of\ input\ signal\ causing\ the\ deviation}$$

The sum of all the sideband power levels plus the carrier power level will equal the total of the transmitter output power. Using special equations, known as Bessel functions, the sideband levels can be calculated.

For example, an FM system with a deviation from the carrier of 3 kHz and an input signal of 1 kHz will have a modulation index of 3. The sidebands and their relative power levels are illustrated in Figure 1.38.

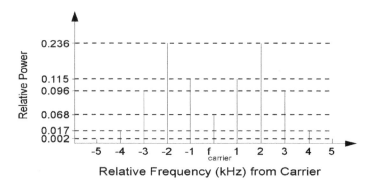

Figure 1.38
Sidebands for FM with 1 kHz signal and 3 kHz deviation

If the deviation of the carrier is 3 kHz and the input signal is 5 kHz, then the modulation index becomes 0.6. Figure 1.39 illustrates the relative power levels of these sidebands.

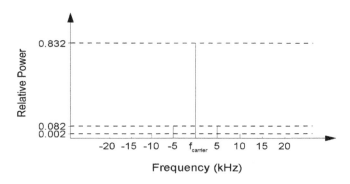

Figure 1.39
Sidebands for FM with 5 kHz signal and 3 kHz deviation

From the above figures, it can be seen that the larger the modulation index, the easier it is to detect the information in the received signal.

Although the sidebands appear relatively small, they may be of sufficient magnitude to cause interference to receivers close to their frequency. Therefore filtering of the distant sidebands is very important.

Because FM systems change RF frequency with a change in the input signal amplitude, they are able to convey dc voltages.

1.6.5 Phase modulation (PM)

Frequency modulation and phase modulation are often referred to as angle modulation schemes. Phase modulation is the process of varying the phase of the carrier wave in sympathy with the information input signal. It should be noted though that frequency and phase modulation are NOT independent of each other, since phase cannot be varied without also varying the frequency.

For PM systems, a true PM receiver will only respond to the rate of change (i.e. instantaneous charges) of the carrier frequency. Therefore, PM cannot directly convey dc levels. The deviation of the carrier is directly proportional to the rate of change of phase and the total amount of phase change of the carrier. This rate of change of phase is directly proportional to the frequency and the rate of change of amplitude of the input-modulating signal. Therefore, with PM systems the deviation increases with both the instantaneous frequency and the amplitude of the modulating signal. To make a comparison, deviation in FM systems is proportional only to amplitude of the modulating system. Besides this, there is no distinguishable difference.

It is found that implementation of the electronics for PM systems can produce slightly more stable systems. Most modern radio equipment uses FM modulation techniques. The majority of telemetry systems will be implemented with FM radio equipment.

The sidebands produced by PM radio systems are the same as for FM radio systems. FM is sometimes referred to as direct FM and PM as indirect FM.

1.6.6 Pre-emphasis and de-emphasis

An analysis of the audio output of an FM radio receiver shows there is a naturally higher noise spectral density at the higher frequencies than at the lower frequencies. This same link has the reverse effect on the audio bandwidth at the receiver output for a flat audio signal put into the transmitter, where at the receiver output, the higher audio frequencies will have less spectral density than the lower frequencies.

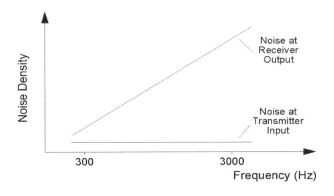

Figure 1.40(a)
Natural noise levels at FM receiver output

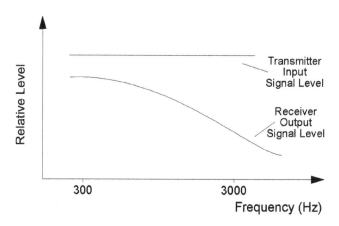

Figure 1.40(b)
Spectral output of an FM receiver for a flat audio input signal at the transmitter

The most commonly used approach to overcoming this inherent problem is to use **pre-emphasis** in the transmitter and **de-emphasis** in the receiver.

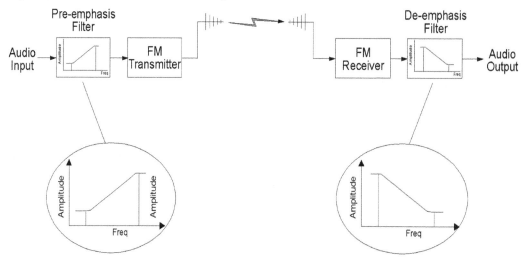

Figure 1.41
Pre-emphasis/de-emphasis circuit

With pre-emphasis, the high frequency components of the input audio are emphasized, prior to modulation in the transmitter and therefore, before noise is introduced into the link. This to some extent will equalize the degradation effect of the link on audio frequencies illustrated in Figure 1.40(b) and provide better use of the available transmission bandwidth. At the receiver, the inverse is performed by de-emphasizing the high frequency components, so that the original flat response of the audio signal is restored. In doing this, the high frequency components of noise are also reduced, thereby effectively increasing the output signal-to-noise ratio.

This process is very effective for voice communications, but will cause distortion to digital signals, in data-over-radio systems.

1.6.7   Comparison of modulation schemes

Figure 1.42 illustrates the difference between the three modulation techniques discussed.

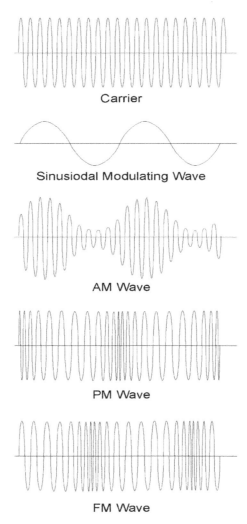

Figure 1.42
Illustrating AM, PM and FM waves produced by a single tone: (a) Carrier wave, (b) Sinusoidal modulating wave, (c) Amplitude modulated wave (d) Phase modulated wave (e) Frequency modulated wave

It is seen that a distinction between an FM signal and a PM signal can only be made when compared to the input-modulating signal. Therefore, a system of transmitters and receivers can be a mix of FM and PM and operate successfully (with a few technical constraints). For the purposes of this book, it would be satisfactory to only refer to FM systems and the characteristics can be considered the same for PM systems.

Angle modulation radio systems provide significantly improved discrimination against noise and interference when compared to amplitude modulation systems. Hence, the majority of radio systems used today in the VHF and UHF bands, are PM/FM.

As a technical note, PM is sometimes preferred to FM. The first reason is that with PM, the input signal can be applied directly to the carrier RF amplifier and does not require modulation of the carrier oscillator, as is required in FM. This makes frequency-stability easier to achieve.

Secondly, a PM system has natural pre-emphasis and de-emphasis characteristics, on the input-modulating signal to the transmitter, and on the output-demodulated signal from the receiver, respectively. This helps to reduce the noise content of the system. However, modern FM systems will use pre-emphasis and de-emphasis circuits, at the transmitter and receiver, to counter this.

Because of the nature of PM, higher deviations can be generated relatively easily when compared to FM systems. In practice, the older type radios that used crystals for the carrier frequency source and straight multipliers after the modulator, used phase modulation. The modern synthesized radios that use phase-locked loops as modulators use **frequency modulation**.

1.7   Amplifier

An **amplifier** is a two-port device that amplifies audio, radio, or intermediate frequencies, from a low level to a higher useable level.

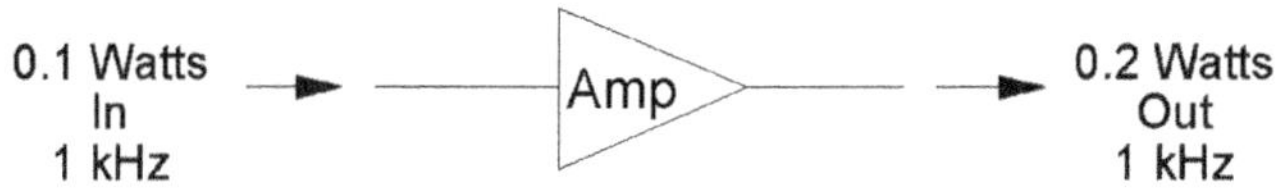

Figure 1.43
Amplifier

Gain for an amplifier is expressed in dB. Therefore the gain for the above amplifier is:

$$10 \log \frac{0.2}{0.1} = \underline{3\,dB}$$

1.8   Power amplifier (PA)

A power amplifier is a two-port device that takes a low level RF signal and amplifies it to suitably high RF levels, for connection to an antenna. For example, a signal of 0.5 watts at 450 MHz is fed into a base station PA and is amplified to 5 watts at 450 MHz. The symbol for a PA is:

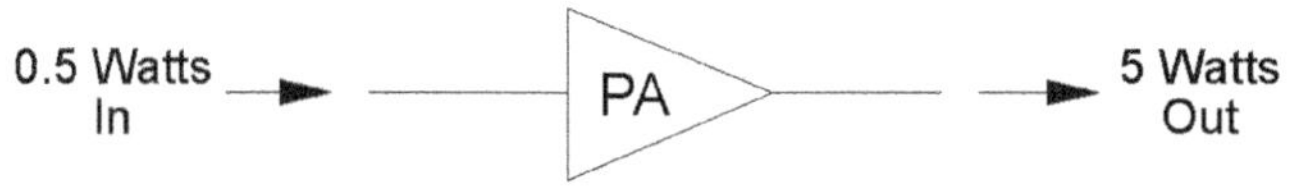

Figure 1.44
Power amplifier

Gain of the amplifier illustrated in Figure 1.44 is 10 dB.

1.9   Oscillator

An **oscillator** is a special type of amplifier where part of the output is fed back to the input, causing the amplifier to oscillate at a specific frequency. An oscillator can be configured to generate any required frequency in the radio frequency band. By filtering the output, a sinewave is produced. The normal illustration of a sinewave oscillator is shown in Figure 1.45.

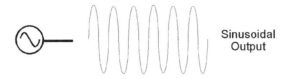

Figure 1.45
A sinusoidal oscillator

An oscillator can be configured to provide only a single output frequency, or variable, where an applied changing voltage will change the oscillator frequency output in sequence.

These are referred to as voltage controlled oscillators (VCO) or variable frequency oscillators (VFO).

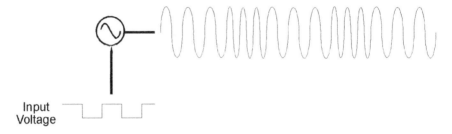

Figure 1.46
Voltage controlled oscillator

The stability of a fixed oscillator will greatly affect the performance of a radio. If the oscillator strays off frequency, the transmitter can cause interference to other nearby channels, while the receiver may not receive the incoming signal correctly (as the frequency the receiver is set to by the drifting oscillator will be different to the received frequency) and will be susceptible to interference from other sources.

Stability of an oscillator is usually measured in percentages or parts per million (PPM). An oscillator with a stability of 0.005% (or 5 PPM) is good, while 0.002% (or 2 PPM) would be excellent.

The agility of a variable frequency oscillator can be important for radio equipment that is required to quickly and easily change frequency, as when a single radio is required to monitor or work on different channels.

The most advanced form of an oscillator is a frequency synthesizer. Here, using digital counters or dividers in a frequency controlled loop, a filtered sinusoidal output voltage is provided, which is tunable with high resolution, over a wide frequency range and yet still has the stability and accuracy of a fixed crystal controlled oscillator. With the use of only one stable frequency source, a whole range of stable frequencies can be obtained. Figure 1.47 shows a simple synthesizer.

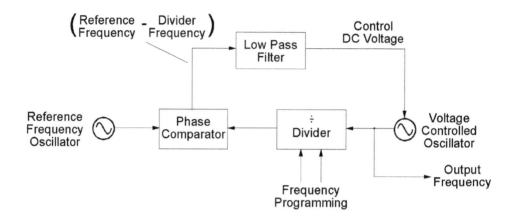

Figure 1.47
Frequency synthesizer/phased locked loop

Examining the circuit from left to right, the phase comparator (also referred to as a detector) mixes the reference frequency with the digital frequency output of the divider, and provides an output equal to the reference frequency – divider frequency, which will be a low audio frequency. This is filtered by the low pass filter and turned into a dc voltage that follows the change in input audio frequency. The dc voltage then controls the VCO providing the required output frequency.

By selecting the required ratio for the divider, very accurate control of the VCO is provided by feedback through the phase comparator. This is the principle of a phased locked loop.

1.10   Filters

Filters are an important tool in radio and are used extensively throughout radio systems. Basically, a filter is a device that selectively removes frequencies from a band of frequencies at its input, and thereby only letting through chosen frequencies to the output.

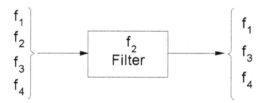

Figure 1.48
A filter (removes frequency f_2)

Figure 1.48 illustrates a filter that is designed to remove frequency f_2.

With today's crowded frequency bands, highly selective filters are required to separate radio systems operating only a few kilohertz apart. Filters will be used in the audio part of a circuit, in the intermediate part of a circuit, in the RF stage of a transmitter and receiver, and then connected between the antenna and the radio equipment (providing extra filtering of the RF signal being transmitted and received). The first three types of filters will be constructed of electronic components, while the latter type will generally be of mechanical construction.

1.10.1   Filter characteristics

Filters will have one of four main characteristics as illustrated below:

Low pass filters (LPF)

An LPF will allow only frequencies below a selected cut off frequency to pass from the input, through to the output.

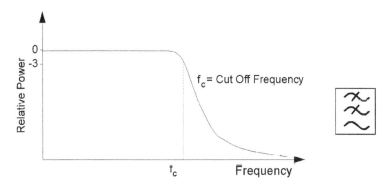

Figure 1.49
Low pass filter

High pass filter (HPF)

An HPF will allow only frequencies above a selected cut off frequency to pass from the input, through to the output.

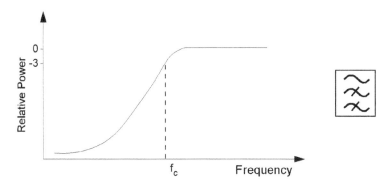

Figure 1.50
High pass filter

Band pass filter (BPF)

A BPF will allow only frequencies between two select cut off frequencies to pass from the input, through to the output.

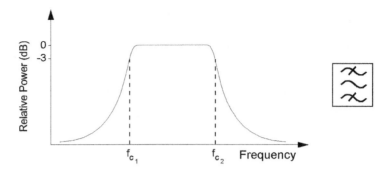

Figure 1.51
Band pass filter

Band reject filter (BRF)

A BRF does NOT allow frequencies between two select cut off frequencies to pass from the input, through to the output.

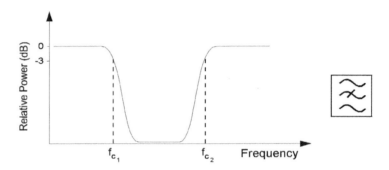

Figure 1.52
Band reject filter

A filter is fundamentally a tuned circuit, where that circuit will resonate at a specific frequency or set of frequencies. For example the single frequency band pass filter, illustrated in Figure 1.53, is a circuit that resonates the most when a frequency of 175 MHz is applied to it.

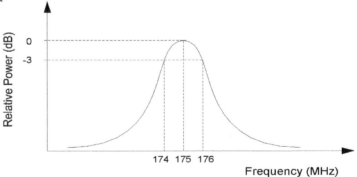

Figure 1.53
Single frequency band pass filter

A close look at the graphs for the filters in Figure 1.54 indicates that the effectiveness of a filter depends on the slope of the attenuation curve, at the cut off frequency. A relative power drop of 0.5 compared to the maximum power is a 3 dB level drop – the steeper the slope, the more effective the filter.

The sharpness of a filter (the steepness of the curve's slope), is measured using a figure referred to as the Q factor. To calculate this factor, it is required to know the bandwidth of the filter. The bandwidth points of measurement are determined, where the relative power of the filter pass or rejection band, has dropped by half, i.e. by 3 dB. This is illustrated in Figure 1.53, where the 3 dB points on the filter response are 174 MHz and 176 MHz, and therefore the bandwidth is 2 MHz.

Q then is defined as:

$$Q = \frac{Frequency\ of\ resonance}{Bandwidth\ of\ filter}$$

For Figure 1.53, the Q would be 87.5.

Figure 1.54 illustrates how varying the Q affects the filter characteristics.

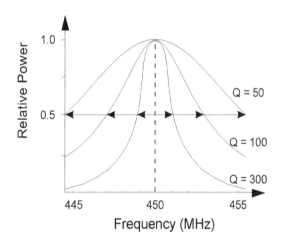

Figure 1.54
The effect of varying Q on a filter

It should be noted that the higher the frequency, the less effective the Q value is. For example, a Q of 300 at 900 MHz represents a filter bandwidth of 3 MHz. At 9 MHz, a Q of 300 represents a filter bandwidth of 30 kHz. In both the VHF and UHF bands, the channel bandwidth allocations are regulated to between 12½–30 kHz, depending on application and frequency band. Therefore, very precise filters with very high Qs are required in high frequency bands.

1.10.2 Types of filters

There are a number of types of filters available. The following section will examine the more common types.

1.10.2.1   LCR passive filters

The most common filter used is that which is constructed of the passive electronic components, inductors, capacitors, and resistors.

Figure 1.55 illustrates a simple low pass filter.

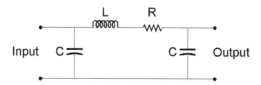

Figure 1.55
Low pass filter using an LRC network

There are an infinite number of variations, on how these components can be combined to create various filter characteristics. Large volumes of material are available on the science of filter design, using passive electronic components.

1.10.2.2   Active filters

Active filters are constructed of powered integrated circuits. These integrated circuits are normally operational amplifiers (referred to as OP-Amps). Active filters are normally used, as they do away with the requirement of having to use cumbersome inductors. Only resistors, capacitors, and OP-Amps are required with active filters, to obtain the same filter characteristics, as using LCR combinations. Figure 1.56 illustrates a simple low pass filter, using active components.

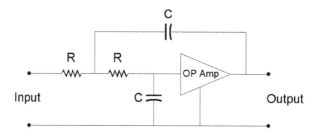

Figure 1.56
Active filter using R, C, and OP-Amp

Again, there are an infinite number of variations, on how these components can be combined, to create various filter characteristics, and large volumes of material exist on the science of designing them.

1.10.2.3   Crystal filters

In a crystal filter the signal is applied to a piece of Piezoelectric quartz crystal, which converts the electrical signal to mechanical vibration. The crystal will have a very sharp resonant frequency at which it will resonate and represent, almost zero-impedance to the signal.

Crystal filters can have Qs up to around 100 000. They are very frequency-stable, but because of mechanical constraints, are limited to relatively low frequencies of up to, approximately, 20 MHz.

Figure 1.57(a) illustrates the LCR equivalent of a crystal and Figure 1.57(b), a band pass filter using crystals.

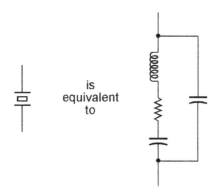

Figure 1.57(a)
LCR equivalent of a crystal

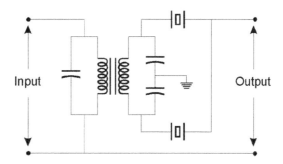

Figure 1.57(b)
A band pass crystal filter

1.10.2.4 Ceramic filters

The same piezoelectric effects that occur in crystal filters also occur in certain ceramics. They are constructed and used in a manner similar to quartz crystals.

Ceramic filters do not have as good a Q, nor are they as stable as quartz filters but are generally cheaper to construct. They will operate up to a frequency of, approximately, 20 MHz.

1.10.2.5 Mechanical filters

A mechanical filter consists of a hollow tube in which metal discs are placed and then coupled, with metal rods, to adjust their position in the tube. Electrical signals are converted to and from mechanical vibrations by transducers on top of the tube. High quality mechanical filters can have Qs up to the 10 000. They are very stable and robust.

Mechanical filters are only effective up to frequencies of around 500 kHz. They are often used with high-power low-frequency transmitters to obtain frequency stability.

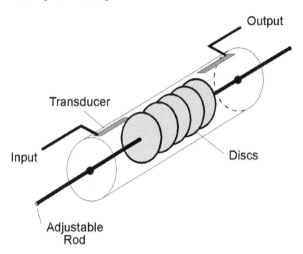

Figure 1.58
Mechanical filter

1.10.2.6 Stripline and coaxial tuned filters

As frequencies move up to and beyond 1 GHz, special filters are required. They normally consist of two small pieces of metal placed close together. At high frequencies, the metal pieces form an equivalent LRC resonant network.

The type of network formed depends on the spacing of the pieces, the thickness, width and length of each piece and the type of material they are made of.

A common implementation of this type of filter consists of a strip of etched copper track on a circuit board, with a copper ground plane on the reverse side. Variable inductors or capacitors are connected to the strips to allow minor adjustment to the resonant frequency. This is referred to as a stripline filter. Figure 1.59 illustrates a stripline filter and its equivalent LRC circuit.

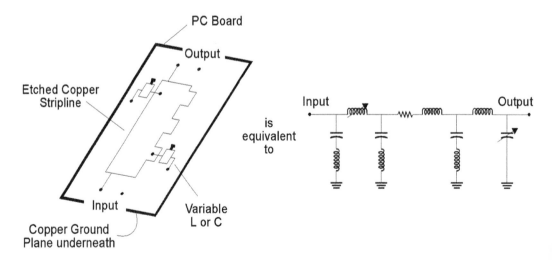

Figure 1.59
Stripline filter and equivalent LCR network

A second filter that operates on the same principle is a coaxial line filter. A piece of coaxial cable will have natural LRC components that vary with frequency. Therefore, short pieces of select coaxial cable can be used to setup tuned resonant circuits, for the required resonant frequency. For this configuration, the outer shield acts as the equivalent ground plane and the inner conductor as the equivalent stripline. Again, variable inductors and capacitors can be used to finely adjust the resonant frequency.

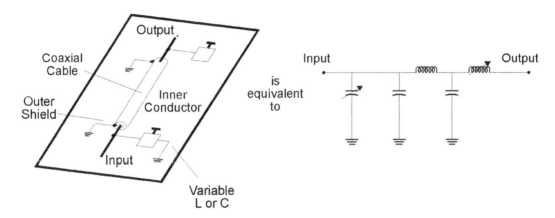

Figure 1.60
A coaxial line filter and its equivalent circuit

The stripline and coaxial filters are extensively used in transmitters and receivers of UHF radio equipment. These circuits can provide Qs of up to 100 000 depending on frequency and construction – the stripline filter having the flexibility to provide higher Qs than the coaxial line filter.

1.10.2.7 Cavity filters

Another common technique is to have two flat discs close together in a hollow tube. This is referred to as a cavity filter. Figure 1.61 illustrates how it is constructed.

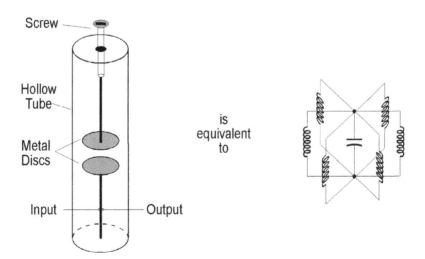

Figure 1.61
The cavity filter and its equivalent circuit

The bottom disc is secured to the bottom of the tube by a long metal rod. The top disc is fixed to an adjustable screw rod. The two discs represent a large capacitor and the surrounding tube, an infinite number of small inductors. These filters can produce very high Qs of 100 000 or more. They are often used in the VHF and UHF frequency bands. A cavity filter at 450 MHz, will be approximately 40 cm high by 15 cm in diameter and in some cases, several may be connected in parallel, to provide higher Qs. Because they are relatively bulky, they are generally only placed on the output stages of transmitter and receivers, before connecting to antennas.

From a project implementation perspective, cavity filters are very important when implementing radio installations for telemetry systems. For this reason, section 1.15 is devoted to discussing the various types of cavity filters and how they are used in radio systems.

1.11 Transmitters

A transmitter is the device that is used to generate RF energy, which it modulates with an information input signal, and then feeds to an antenna to be conveyed to one or more receivers.

A brief overview will now be provided of the fundamental structure of AM and FM transmitters.

1.11.1 AM transmitters

Figure 1.62 illustrates in block form, the construction of a simple SSB-AM transmitter.

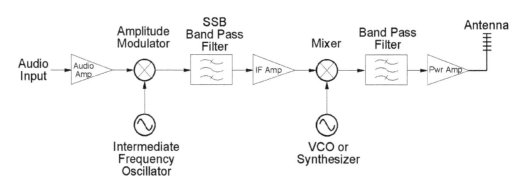

Figure 1.62
SSB-AM transmitter

To explain the operation of the transmitter, follow the figure from left to right. Firstly, the audio amplifier increases the audio input to a level suitable for application to the amplitude modulator. The modulator produces a signal consisting of a carrier and two sidebands. The carrier frequency in this case would be an intermediate frequency, which is equal to the oscillator frequency, at the input to the modulator. (This intermediate frequency is normally significantly lower than the final RF frequency.) The two sidebands will correspond to the audio input frequency.

Intermediate frequency (IF) is the term used to describe the frequency used in radios, for easier manipulation by the transmitter electronics that lies between the input audio frequency and the final output radio frequency.

A band pass filter is then used to filter out all frequencies, except for one sideband. The output is fed to an amplifier, which increases its level suitable for application to a mixer.

The mixer is a device similar to a modulator that converts a low level signal (in this case, the IF sideband frequencies), from one frequency to another, by combining it with a high level signal from an oscillator. The mixer is generally a non-linear device and will produce a number of frequencies, at its output. These consist mostly of the sum and difference frequencies, of the IF and oscillator frequencies.

At the final stage, normally, the sum of the two frequencies is filtered out by a band pass filter and is applied to a power amplifier for transmission to an antenna.

1.11.2   PM and FM transmitters

Figure 1.63 illustrates, in block form, the fundamental construction of an FM and a PM transmitter.

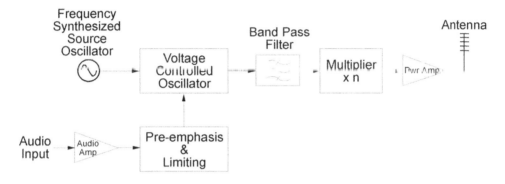

FM Transmitter

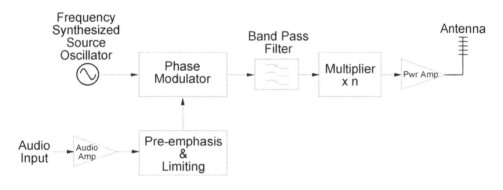

PM Transmitter

Figure 1.63
Block diagrams for FM and PM transmitters

Following the FM transmitter figure from left to right, a stable frequency synthesizer forms the frequency source for a voltage-controlled oscillator.

The audio frequency input is fed through an audio amplifier to a unit that carries out any required processing on the audio. The unit, firstly, limits the maximum amplitude of the audio signal, so that the transmitter does not over-deviate when fed with high amplitude audio signals. (Over-deviation would cause RF distortion and significant interference to nearby channels.) The audio processing unit then provides signal conditioning, referred to as pre-emphasis, to decrease the effect of noise on the final RF signal. Pre-emphasis and de-emphasis are discussed in detail, in section 1.6.6.

The modified audio signal is then applied to the VCO to produce a frequency-modulated signal. This signal is the intermediate frequency as it is not yet at the higher RF frequency.

Following the VCO, the signal is fed to a band pass filter that will filter out any spurious signals produced by the synthesizer and the VCO. The signal is then frequency multiplied up to the required RF frequency. For example, in a UHF radio the IF frequency may be 20 MHz and it would be multiplied up to 450 MHz. The signal is then fed to a power amplifier, to increase it to a level suitable for transmission to the antenna.

The PM transmitter is virtually the same as the FM transmitter, except that the IF frequency is fed directly to a phase modulator, instead of varying the output of a voltage controlled oscillator. This system provides better frequency stability and accuracy, as it is not dependant upon a linear variable frequency source.

1.12 Receivers

A receiver is a device used to detect radio frequency waves at a selected frequency that are fed to it from an antenna, and then to decode these into an intelligible audio signal. A brief overview will now be provided for the fundamental structure of receivers.

1.12.1 AM receiver

Figure 1.64 illustrates the structure of a simple SSB-AM receiver.

This type of receiver is referred to as a superheterodyne receiver. Heterodyning refers to the process of mixing two or more frequencies together to produce other different frequencies. (Normally the produced frequencies are the sum and difference of the mixed frequencies.)

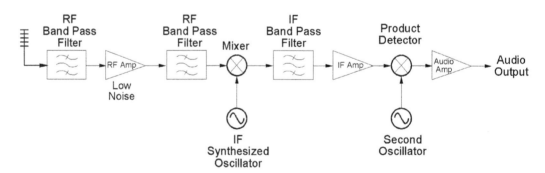

Figure 1.64
AM-SSB Superheterodyne receiver

With reference to Figure 1.64 and moving from left to right, the RF signal is fed to the *front end* of the receiver from the antenna and is selectively detected by a series of filters and amplifiers. All receivers require very narrow (high Q) band pass filters to select only the frequency on which the receiver is to operate. If the filters are not of a 'high-selective' quality and the amplifiers of very 'low-noise' quality, the other frequencies, close to the operating frequency, will be let through to the receiver and will cause interference problems.

The signal is then fed through to a mixer where it is mixed with an IF oscillator frequency. The required IF sum or difference frequency, is then filtered by a band pass filter from the mixer. The chosen IF signal is then amplified to a level suitable for applying to a product detector.

The product detector is an AM demodulator. The IF frequency and the second oscillator frequency will be very close and the product detector will produce an audio frequency output that corresponds to the audio put into the transmitter input.

1.12.2   FM and PM receivers

Demodulation of an FM signal is the same process as demodulation of a PM signal. Therefore, the FM and PM receivers have the same fundamental construction.

Figure 1.65 illustrates an FM receiver in block form. When compared to the AM receiver, it can be seen that there is minimal fundamental difference between them. The only difference being at the stage after the IF amplifier, where an AM product detector is replaced with an FM discriminator or a phased locked loop.

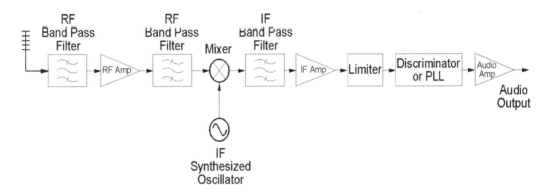

Figure 1.65
FM/PM superheterodyne receiver

The FM discriminator has a response, as illustrated in Figure 1.66. There is a linear change in the output voltage of the discriminator as the frequency deviates from the carrier frequency (in this case the equivalent IF frequency). This changing dc voltage would then be following the audio input that was used to modulate the IF at the transmitter, i.e. it is demodulating the FM or PM signal.

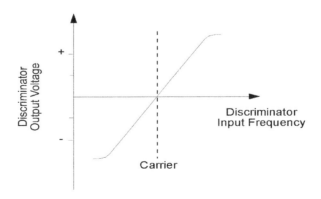

Figure 1.66
Discriminator characteristic

When the RF frequency is equal to the carrier frequency (i.e. no deviation), then the output of the discriminator is 0 volts.

The second method of FM demodulation is to use a phase locked loop. Here, the IF frequency is used as the frequency source and the output comes from the filter line feeding into the voltage-controlled oscillator. This is illustrated in Figure 1.67. This is the most common method of demodulation used in modern FM radios.

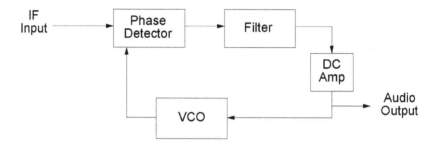

Figure 1.67
A phased locked loop used as a demodulator

1.13   Antennas

1.13.1   Introduction

Antennas come in a bewildering array of shapes and sizes and one might be tempted to think that almost any old bit of aluminum will do. In the early days of television, some people used old bicycle wheels as antennas and swore by them!

In fact, antennas are generally precisely engineered and must be carefully specified and installed, if they are to perform properly. Even the humble television antenna, which is cost-reduced to an extreme degree, is capable of good results if treated kindly.

1.13.2   Theory of operation

All antennas do exactly the same job. If they are transmitting antennas, they convert electrical energy into electromagnetic energy and radiate or launch this energy into free space. If they are receiving antennas, they capture the energy waves in free space and convert then into electrical energy. A single antenna generally performs both functions at the same time.

The first requirement for an antenna is to provide the gain and directivity required by the system. In a point-to-point link, it is vital that all the transmitter power is pointed towards the receiving end and similarly at the receiving end that the maximum signal is gathered from the transmitting end. To understand how an antenna works, it may be helpful to delve into a little bit of theory. A point source of light, similar to a torch bulb but without any base, would radiate light equally in all directions and the light pattern would be in the shape of a ball. In a darkened room, the light might be strong enough to allow a book to be read and of course, the distance from the light would always be the same.

In electrical terms we could say that the lamp was an isotropic source with a gain of 0 dB. If we now placed a mirror close to the light, one side of the room would be dark and the illumination on the other side would be almost twice as strong, representing a gain in electrical terms of 3 dB, over the isotropic lamp (i.e. twice the power in the opposite direction). Obviously, our book could be illuminated at a greater distance from the lamp. If it were necessary to read still further from the lamp, it could be put into the focal point of a parabolic reflector, which is exactly the shape of the reflector in a good spotlight. The lamp might now have a gain to 30 dB over the isotropic lamp and it would be easy to read the book outside the room and down at the end of the corridor!

From the analogy above, it can be seen that the gain in illumination in the required direction and the directivity of the light beam are really the same, so that the narrower the beam-width, the greater the gain in illumination.

Radio antennas behave in a similar manner to the mirror and reflector analogy, used above. The lamp is replaced by an element, usually a dipole of some type and often called the illuminator, and the reflector is called just that. In the case of antennas used above 400 MHz, a dish antenna looks very much like an unpolished searchlight mirror.

An isotropic source is a theoretical concept and we cannot manufacture a perfect point source of light because the filament has a certain size and the connecting wires will get in the way of some of the light. In the same way it is not possible to make an isotropic radio illuminator, so a dipole is used as a practical reference standard. Because a dipole distorts the spherical radiation pattern, it actually provides a signal gain of 2.14 dB over the isotropic source.

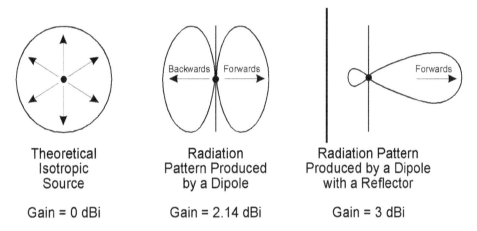

Figure 1.68
Some simple radiation patterns

The second important requirement of an antenna is to provide a correct electrical termination for the transmitter/receiver and transmission line that feeds it. A plumbing analogy can be used here. Imagine a pump transferring water to a distant tank through a pipeline. If the pump has a 10 cm outlet flange, the designer should obviously use a 10 cm pipe to run to the tank. If the tank had a 7 cm inlet flange a high back pressure would be created and the maximum transfer of fluid to the tank would not be achieved. Whilst the analogy breaks down if a 13 cm pipe were used, it serves to illustrate what is known as the maximum power transfer rule, which simply states that to achieve maximum power transfer from a source to a load (i.e. an antenna), the impedance of the source, the transmission line and the load shall all be equal.

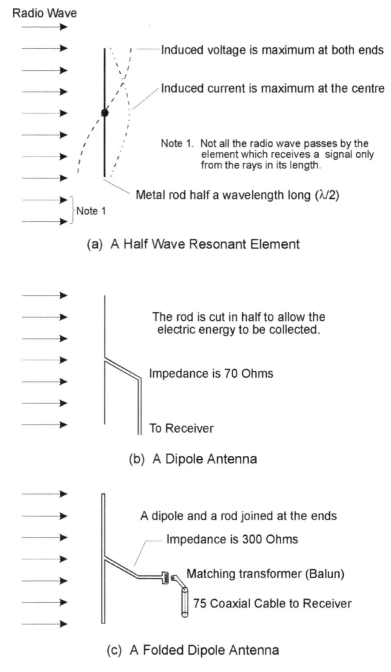

Figure 1.69
How a simple dipole works

Figure 1.69 illustrates the fundamental operation of a dipole. The incoming electromagnetic waves induce electron flow in the dipole element. Figure (a) illustrates the voltage and current magnitudes in the dipole element. As the electrons approach the ends of the element, they have the highest potential but the least speed; that is, the highest voltage and least current.

The reverse is true in the middle of the dipole. Therefore, impedance of the dipole is highest at the ends and lowest in the middle. Because it is best to tap-in at the lowest impedance (so only a low impedance source is required to match the load), the coaxial feed connection is in the middle of the element. This has a natural impedance of around 70 ohms.

The most common form of dipole element is the *folded dipole,* illustrated in figure (c). This has a natural impedance of around 300 ohms. Matching transformers (Baluns) are used to connect to 50 and 75 ohm coaxial cables.

As was discussed earlier in the chapter, the radio wave that is emanating from the element is constructed of an electrical and a magnetic field (refer to section 1.2). The electric field is parallel to the element. The orientation of the electric field is referred to as the *'polarization'*. Therefore, if the elements of an antenna are straight up and down, the antenna is vertically polarized and if they are lying flat, it is horizontally polarized. It is also possible to have circularly polarized radiation patterns by using circular elements.

It would be ideal if an antenna could direct the radio wave into a parallel sided beam, rather like a laser beam, but this is not so and the beam is formed onto an arc, which becomes wider as the distance increases, exactly like a light beam from a torch. If a torch is pointed onto a large wall, the center of the beam will be a bright spot, which gradually fades as we look away from the center. In the case of a radio wave, the beam-width is defined as the arc formed at the points where the received signal strength has fallen to half power or minus 3 dB.

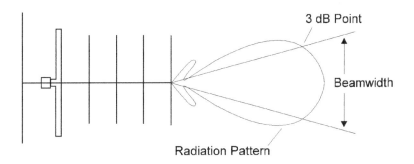

Figure 1.70
3 dB beam-width of an antenna

1.13.3 Types of antennas

Antennas can broadly be divided into four types:

- High frequency
- Single or stacked dipole
- Resonant element or Yagi
- Parabolic dishes

The first three are used a lot in radio, and the parabolic dish is of most relevance to microwave and satellite systems.

High frequency antennas operate up to frequencies of about 30 MHz and they usually consist of long wire sections carefully designed to provide the correct impedance match, which is usually 600 ohms and installed to point in the desired direction. Whilst some types, known as log periodic antennas, are able to be rotated, others such as rhombic antennas occupy as much as 6 Hectares (15 Acres) and are very much a permanent installation.

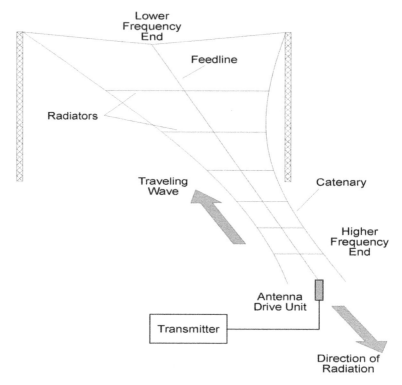

Figure 1.71
Illustration of a log periodic

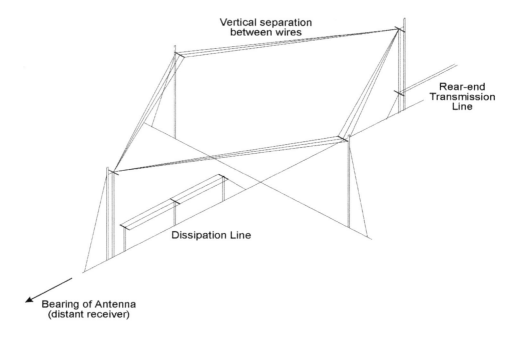

Figure 1.72
Illustration of a rhombic

The simplest antenna is a straight dipole. A single piece of thin metal is cut to ⅛, ¼, ½, or 1 times the wavelength of the operating frequency. The effective wavelength and directivity is sometimes extended using a coiled piece of metal. By stacking several dipoles on top of each other and feeding them in phases, it is possible to provide increased gain and directivity. This type of antenna is sometimes referred to as co-linear. They can provide 3 to 6 dB gain. These are often used at master sites in point-to-multipoint systems because they are omni-directional.

Resonant element or Yagi antennas (the name comes from the Japanese Dr Yagimoto, who invented them) are used in the important VHF and UHF bands centered on 80, 160 and 450 MHz. These are the antennas widely used for broadcast-television and are seen on many a rooftop. They consist of a driven element or illuminator, which is usually a dipole or a folded dipole, with one or more reflectors mounted behind the driven element and from one to ten or more director elements in front of the driven element; the more the number of elements, the higher the gain and the directivity. Typically, a Yagi operating in the 160 MHz band with 10 elements will have a gain of 13 dB. The antenna impedance is usually 50 ohms.

Because the length of the elements is inversely proportional to the frequency, an 80 MHz Yagi element will be about 1.9 m long and in this band, antennas of more than four elements become very cumbersome, whereas at 450 MHz, the elements will be only 330 cm and it is not uncommon to specify 15 element antennas.

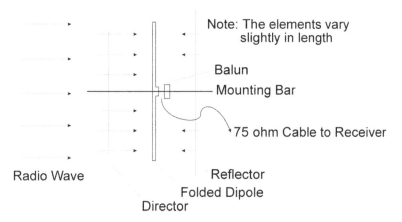

Figure 1.73
How a simple Yagi antenna works

1.13.4   Antenna installation

There are a number of important considerations that should be taken into account when selecting and installing antennas. These are summarized below.

- The antenna should be selected according to:
    - The required frequency of operation
    - Radiation pattern; which affects the coverage area
    - Spectrum Management Agency regulations
- The antennas should be positioned at sufficient distance from neighboring antennas, both vertically and horizontally to avoid radiation pattern distortion.
- The antenna should be correctly positioned on the leg of the mast. The horizontal distance it is placed from the mast leg is critical in determining the antenna radiation pattern, as the leg acts as a reflector element.
- Antennas should be as broadband as possible, so multiple transmitters, and receivers can be multicoupled onto one antenna.

- Antennas should not be constructed of dissimilar metals, or corrosion occurs between the different parts and a diode effect is established, causing the production of intermodulation products (refer to section 1.15).
- The wind loading effect of the new antenna should be calculated and taken into account on the mast.
- The receiving antenna, is normally located at the highest point on the mast, to achieve maximum receive signal strength and the transmitter antenna at a suitably lower level, to achieve the required isolation.

1.13.5   Stacked arrays

If the gain of a single antenna is not sufficient, two or four identical antennas can be connected in parallel or stacked. When this is done, the transmission line is connected to an impedance matching transformer with two or four outlets to feed each antenna. A two-stack array has a gain a little less than 3 dB more than that of a single antenna and a four-stack array, a little less than 6 dB more than the single antenna.

There are a number of important rules to be observed when stacking antennas:

- The cables from the impedance matching transformer must all be exactly the same length, even if surplus cable needs to be coiled up.
- The antennas may be stacked vertically, horizontally or in a square pattern but they must all be in the same vertical plane. That is to say that the driven elements of all antennas must be exactly the same distance from the receiving station.
- All the antennas must be installed in the same phase. If the first antenna is installed with its connector on the lower side, then all the others must be installed the same way.

If these rules are not observed there will be a phase difference in the wavefronts from each antenna and the stacking gain will not be achieved. If, in a two-stack array, one antenna was installed with the connector facing up and the other with the connector's face down, the wavefronts from the two antennas would be 180° out of phase. The resultant combined reading would be 30 dB lower as opposed to an expectant gain of 3 dB per antenna.

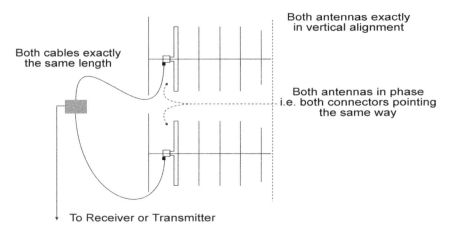

Figure 1.74
A two-stack Yagi antenna

1.14   Cabling

Coaxial cables are used, almost without exception, for all antennas operating between the HF-band, of frequencies up to the SHF band, around 2 GHz, when waveguides begin to take over.

The impedance of a coaxial cable is determined by the diameter of the inner conductor and the spacing, between it and the surrounding shield. Although the characteristic impedance of a television antenna is 75 ohms, most communications antennas have an impedance of 50 ohms and care should always be taken to use the correct cable.

The size of a coaxial cable is determined by two conditions – the transmitter power being fed to the antenna system and the frequency to be used.

If a transmitter had an output power of 500 watts, the peak voltage across a 50 ohm cable will be 223 volts, and the current will be about 3.3 amps. If the dielectric insulation is insufficient, breakdown of the cable will take place and if the inner conductor is too small, there will be a high resistive loss in the cable.

Radio frequency energy tends to travel on the surface of a conductor rather than through the center, so a small diameter inner conductor will obviously have a small surface area and consequently high resistance. Thus, as the frequency increases, so should the diameter of the inner conductor, but the impedance of the cable is determined in part by the capacitance between the inner conductor and the screen. Therefore, in order to maintain correct impedance, the size of the inner conductor and the spacing between the conductors, i.e. the dielectric, are the critical design elements to be considered.

Smaller types of coaxial cable, up to about 10 mm diameter, use a copper braided sleeve as the outer conductor and this is efficient and cheap to manufacture. The largest coaxial cables can be up to 200 mm or 8 inches in diameter and as the cable has to be curved around a bend and as any deformity will badly affect the performance of the cable, a new type of shield conductor was developed.

The first attempt to make a suitable outer-conductor used an aluminum tube, but this was extremely difficult to handle and was soon replaced with a copper tube, which looks like the corrugations of a water tank, but is produced with a spiral corrugation so that the diameter at any point along the length of the tube is always constant. In this way, the average distance between the inner conductor and the outer is always the same so that the impedance remains constant, but the cable can be easily curved around quite a small radius without any damage.

Cable manufacturers publish accurate data on the characteristics of the cables they make. At first glance, the selection of a cable may seem little more than locating the cheapest in the market that is rated to carry the desired power. However, in most cases involving radio links, cost is a minor consideration compared to the priority that must be given to attenuation of the cable.

In general, the larger the cable, the lower will the attenuation be, but the cost and the wind loading of the cable will be higher.

In the design of a point-to-point system, the limiting factor is the total attenuation between the transmitter output and the distant receiver input and this will be discussed later in Chapter 2. Unless the path is a short one, the designer must generally make sure that the attenuation of the antenna feeders is kept low. Therefore, the selection of a coaxial cable will be a compromise between cost and possibly wind-loading, on one hand, and attenuation, on the other.

At frequencies over 2 GHz, the inductive and capacitive losses, in conventional coaxial cables, begin to severely attenuate the transmitted and received signals. Larger cables, and the use of an air dielectric, reduce these losses, but at frequencies over 4 GHz, coaxial cables tend to become impractical and waveguide systems are used.

1.14.1 Leaky coaxial cable

Sometimes it is necessary to provide mobile radio coverage, in tunnels or mineshafts, where it is very difficult to obtain good radio transmission and sometimes, large steel framed buildings cause problems. A special type of coaxial cable was developed to cater to these situations and it looks very much like a normal elliptical coaxial cable that has been cut along the length to remove about 8% of the shield. The result is a series of small windows in the outer conductor through which radiation can occur. The cable, in fact, functions like a long antenna, which acts equally well as a transmitting and a receiving antenna. The attenuation of this cable is higher than that of a normal cable of the same size and it must be terminated at the end with an antenna or a load resistor.

1.15 Intermodulation and how to prevent it, using duplexers, multicouplers, circulators, isolators, splitters, and pre-amplifiers

1.15.1 Introduction

Besides noise and interference that emanate from man-made sources, (cars, electrical motors, switches, rectifiers, etc), there are three other main causes of RF interference. These are produced from other radio equipment. The first and the most obvious source, is another radio user operating close by, on the same frequency, as the system suffering from interference. Unfortunately, besides using special coding techniques to minimize the problem (discussed in section 1.21 – CTCSS) there is little that can be done, short of complaining to the regulatory Government body that issues licenses, or finding out who it is and asking them to stop transmitting.

The second source of interference comes from noisy transmitters that emit spurious frequencies outside their allocated bandwidth. These spurious emissions will tend to fall on other users' channel bandwidths and cause interference problems. Aging transmitters and those that are not well maintained, are normally the culprits.

The third source of interference is known as intermodulation. This is normally the most common source of interference and generally, the most difficult to locate and the most costly to eliminate. The following section will examine this phenomenon in more detail.

1.15.2 Intermodulation

Intermodulation occurs where two or more frequencies interact in a non-linear device, such as a transmitter, receiver, or their environs, or on a rusty, bolted joint, acting as an RF diode, to produce one or more additional frequencies that can potentially cause interference to other users. When two electromagnetic waves meet and intermodulate in a non-linear device, they produce a minimum of two new frequencies – one being the sum of the frequencies and the other being the difference of the frequencies.

A nearby receiver may be on or close to one of the intermodulation frequency products, receive it as noise and interference and then could retransmit it as further noise and interference. For example, if two frequencies a and b interact then they will produce two new frequencies c and d where $a + b = c$ and $a - b = d$. c and d are referred to as intermodulation products.

Of course c and d will be of significantly less magnitude than a and b and their exact magnitude depends on the magnitude of a and b at the point a and b meet, and on the efficiency of the non-linear device at which the intermodulation takes place.

Fortunately this problem is only significant when the two transmitters for a and b are within close proximity. Nevertheless, consideration should be given to intermodulation products produced at a distant location, as these have been known to cause noticeable background noise.

If there are more than two frequencies at one location then the number of intermodulation products possible increases dramatically.

For example, if there are transmitters on frequencies a, b and c at one location then the intermodulation products become –

$$a + b = f_1$$
$$a - b = f_2$$
$$b + c = f_3$$
$$b - c = f_4$$
$$a + c = f_5$$
$$a - c = f_6$$
$$a + b + c = f_7$$
$$a + b - c = f_8$$
$$a - b + c = f_9$$
$$a - b - c = f_{10}$$

This illustrates that the number of potential intermodulation products becomes prodigious as the number of frequencies increases. Unfortunately, the scenario gets worse. Each frequency from a transmitter will produce a significant harmonic at twice, three times, four times, etc its carrier frequency (this is particularly true with FM systems). Each sequential harmonic will be of a lesser magnitude than the previous one.

Therefore, if the transmitter is operating on frequency a then harmonics will be produced at $2a$, $3a$, $4a$, etc. The $2a$ and $3a$ harmonics can be quite large. These harmonics are produced because of resonant properties of antennas, cables, buildings, and tuned circuits in the receivers and transmitters themselves, and due to the harmonic sidebands produced in FM.

Taking these harmonics into account, intermodulation products such as:

(i) $2a - b$

and

(ii) $3b - 2c$

are examples of what can be produced. (i) Is referred to as a third order product and (ii), as a fifth order product. This refers to the total of multiples of each frequency.

These intermodulation products can cause severe interference in radio systems that are located at the same site. Considering these, there are numerous permutations of a small number of base frequencies that may cause intermodulation interference.

Generally, intermodulation products greater than the fifth order are too small to be of consequence and generally, radio systems are engineered only to take into account possible intermodulation distortion up to and including fifth order products. At sites where there are many sensitive receivers on different channels in one building, then calculations to the seventh order may be carried out.

For transmitters that are 1 km or more apart, the intermodulation products are, generally, so small as to not be a problem. The exception may be for first or second order products.

If a system is experiencing unexplainable interference problems, a check should be made for distant intermodulation products.

Another source of harmonics sometimes comes from old buildings or masts, where rusty bolts or nails act as RF diodes (non-linear devices), to produce intermodulation products. These are then detected by nearby receivers and may be retransmitted as noise and interference.

A number of devices have been developed to assist in preventing the formation of intermodulation products and to prevent these products (spurious transmissions and harmonics), from causing interference to nearby receivers or transmitters. All these devices are connected between the transmitter and the antenna. The following sections provide details on these devices and how they are used to eliminate interference problems.

1.15.3   Circulators and isolators

Circulators and isolators are devices made of ferrite compounds that, due to resonant magnetic effects, allow RF energy to flow in one direction only.

Circulators are generally three or four port devices. RF energy flowing into one port will flow in one direction only. Therefore, with reference to Figure 1.75 RF energy entering port A will flow to port B, energy entering port B will flow to port C and so on. RF energy will not flow from port C to port B or from port B to port A.

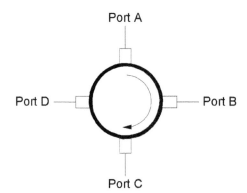

Figure 1.75
A four-port circulator

The characteristic impedance of a port is normally 50 ohms. Therefore, if the port is not terminated correctly, energy will reflect back up the port and on to the next port. For example if port B were left open circuit, energy into port A would travel to port B, be reflected back out again and on to port *C*.

If a 50 ohm terminator is placed on B, then all energy will be absorbed at port B and none will reflect on into port C.

Circulators have a signal loss in the forward direction of approximately 0.5 dB. For signals traveling in the reverse direction, there is about 30 dB attenuation (i.e. an isolation of 30 dB).

Circulators are often used to connect a transmitter to an antenna in conjunction with other filter equipment. This allows RF energy at port A, to flow from the transmitter, to the antenna at port B. At the same time, it prevents unwanted RF energy from flowing back from the antenna at port B to port A. This also prevents intermodulation products from being produced in the transmitter at port A.

This configuration of the circulator, illustrated in Figure 1.76, is often referred to as an isolator.

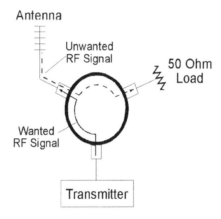

Figure 1.76
A three-port circulator setup as an isolator to connect a transmitter to an antenna

In some cases, there are coaxial cables of other transmitters running parallel up a mast to the main transmitter. These may be from antennas that are situated close to the main antenna. RF energy produced from such cables, or from other stray sources such as mobile radios, will be coupled from the antenna port of the isolator, to the 50 ohm load port and will get absorbed. This is illustrated in the configuration stated above. Therefore, the unwanted coupled signal is isolated from the transmitter.

Some UHF isolators and circulators are constructed from short pieces of tuned coaxial cable, connected between ports. At the correct frequency, the waves are 180 out of phase on the predecessor port and cancel out.

1.15.4 Multicoupler and cavity filters

Multicouplers are, basically, tuned cavity band pass filters (refer to section 1.10.2.7), used for filtering spurious signals from transmitters and for connecting multiple transmitters to one antenna. Figure 1.77 illustrates how the connections are made.

Multicouplers are sometimes referred to as combiners, when used for connecting multiple transmitters to one antenna.

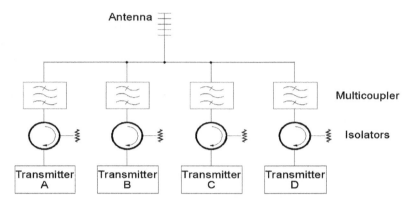

Figure 1.77
Multicoupler used to connect multiple transmitters to a single antenna

In this example, each cavity filter is tuned so that only the frequency of the transmitter, to which it is attached, is passed to the antenna.

In this way transmitter *B* cannot feed back into *A*, nor *C* into *A*, nor any transmitter back into any other transmitter, so long as each transmitter frequency does not fall into the bandwidth of any of the other filters.

Sometimes several cavity filters are connected in series to provide a very high *Q* filter for a transmitter in a noisy RF area. This is illustrated in Figure 1.78(a).

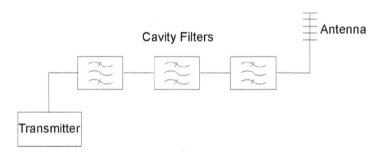

Figure 1.78(a)
Connection of several cavity filters in series

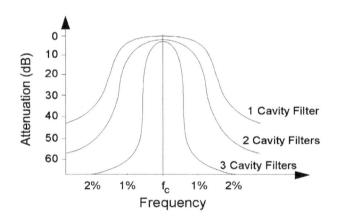

Figure 1.78(b)
Responses of cavity filters connected in series

Figure 1.78(b) illustrates the different responses when connecting cavity filters in series. A manufacture will often quote filter responses as being 'so many dB down' for 1%, 2%, or 3% 'off' the carrier center frequency. For example, a filter operating at 450 MHz may have three cavities in series, with an attenuation of 70 dB at ±1% off the carrier frequency i.e. at 445.5 and 454.5 MHz. Normally a multicoupler or series cavity filters, will introduce between 1 and 3 dB loss into the transmission path.

A series cavity filter can also be placed in series with the receiver to protect it against stray RF from other transmitters. This can also be achieved through a duplexer, which is discussed in the next section. Often on sites where there are a number of transmitters close to a receiver, an extra series cavity filter is required to help prevent serious interference, and intermodulation problems in the receiver. Desensitization and blocking are two other phenomena that occur when transmitters are located close to receivers and can be prevented with extra cavity filters (discussed in section 1.21).

Before purchasing multicouplers or cavity filters, ensure a thorough check has been performed on intermodulation products and that the filtering equipment to be installed, will remove them.

On sites where there are many transmitters, group those together onto one antenna that are relatively close in frequency. This will help keep costs down. In addition, channel separation for each cavity in a multicoupler should be kept to a minimum of 300 kHz in 400–520 MHz band, and 500 kHz in the 800–950 MHz band. Remember, multicouplers are expensive and it is best to engineer the system correctly the very first time.

1.15.5   Duplexer

A duplexer is a device that allows a transmitter and a receiver to be connected onto a single antenna, and operate simultaneously without affecting each other's performance. There are a number of types of duplexers available.

The simplest method is to place a band reject filter in series with the receiver, to reject and absorb the transmitter frequency, as it feeds into the receiver.

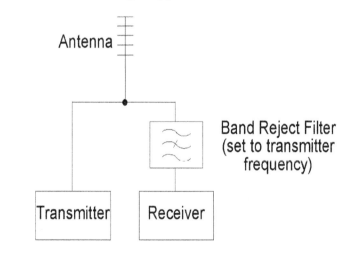

Figure 1.79(a)
Duplexer configuration using a band reject filter

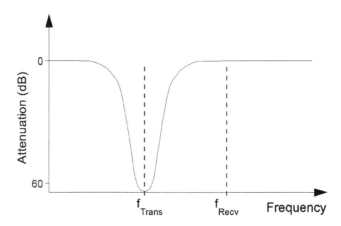

Figure 1.79(b)
Frequency response at input to receiver if a band reject duplexer is used

Figure 1.79 illustrates how this duplexer is configured, and the response at the receiver input.

The main disadvantage with this configuration is that it does not reject any spurious or intermodulation frequencies from the transmitter or other sources that may be on or near the receiver frequency.

A slightly improved duplexer configuration is to use two band reject filters. One is placed in series with the receiver, as in the previous case, and another in series with the transmitter, and is tuned to reject the receiver frequency. Therefore any spurious or intermodulation frequencies close to the receiver frequency that are produced by the transmitter, will be absorbed and will not feed into the receiver. This is illustrated in Figure 1.80.

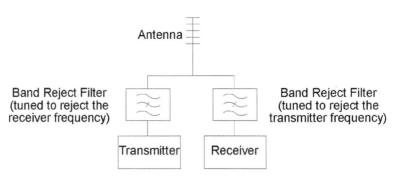

Figure 1.80
Duplexer configuration using two band reject filters

In general, band reject duplexers are cheaper and easier to install than band pass duplexers (discussed next). Nevertheless, a major disadvantage is that since they only selectively reject certain frequencies, all other noise or interference frequencies outside these bands, will be transmitted to air, or received by the receiver, and will cause interference or intermodulation problems.

Sites where there are other transmitters or potential sources of RF noise, band pass filters are generally used.

The most common band pass type duplexer is one, where separate filters are placed in series with the transmitter and receiver. The transmitter filter is tuned to the transmitter frequency so that only frequencies on the transmit frequency can pass to or from the transmitter, and the receiver filter is tuned to the receiver frequency, so that only frequencies on the receiver frequency can pass to or from the receiver.

This is illustrated in Figure 1.81.

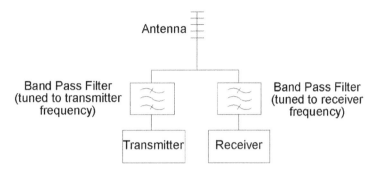

Figure 1.81
Duplexer using band pass filters

It is possible to have a duplexer configuration with a filter on the receiver side of the equipment only. This is not a recommended configuration, as serious intermodulation problems can still occur at the transmitter. Generally, duplexers have two or three series cavities connected on each side.

Characteristics of duplexers vary considerably depending on type and quality used. Some typical parameters are:

For a three cavity filter on the transmit side and on the receiver side of radio for 170 or 450 MHz bands:

- TX–RX isolation of 60–70 dB
- TX insertion loss 2 dB
- RX insertion loss 2 dB

Manufacturers will rate their duplexers for a minimum frequency separation between transmitter and receiver (for example, 4.5 MHz at 450 MHz). If the equipment is operated at closer spacing than recommended then there will be inadequate isolation between the transmitter and receiver, which may cause interference and intermodulation, and high insertion loss of the transmitter and receiver frequencies.

It is also important to select a cavity with correct power handling capability. If the cavity is underrated, it will overheat and detune.

1.15.6 Splitters

A splitter is a simple device used to allow a number of receivers to be connected on to one antenna. They normally consist of a simple resistor network designed so that each port of the splitter has the same characteristic impedance (normally the impedance being 50 ohms).

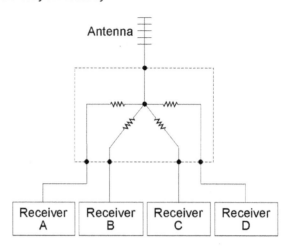

Figure 1.82
A splitter used to connect multiple receivers to one antenna

The major disadvantage with using straight splitter networks is that the power from the antenna is split over the number of receiver ports. In the case illustrated in Figure 1.82, only one quarter of the power coming from the antenna is found at each port, i.e. there is a 6 dB reduction on the received signal.

1.15.7   Receiver pre-amplifiers

A receiver pre-amplifier is a splitter, with a low-noise RF wide-band amplifier, connected to the antenna input. This boosts the received RF input signal from the antenna and overcomes losses introduced by the splitter network.

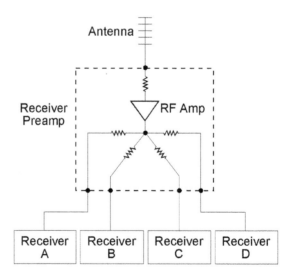

Figure 1.83
Receiver pre-amplifier for connecting multiple receivers to one antenna

A typical receiver pre-amplifier will provide 2 to 4 dB gain before splitting the signal to a number of receivers. The pre-amplifier will also introduce a small amount of additional noise.

1.15.8 Typical configuration

Figure 1.84 illustrates a complete RF filtering assembly that may be typical of a large radio installation.

Note that manufacturers will rate their equipment for a maximum allowed transmitter power. When designing and implementing a system, this power limit should not be exceeded or permanent damage could occur to the isolators and multicouplers.

Besides the elimination of intermodulation problems the filtering equipment described above is used for a number of other reasons including:

- To minimize the cost of antennas and feeders
- To optimize the antenna height for all users operating from one site
- To minimize antenna radiation pattern distortion from too many antennas being too close on the mast

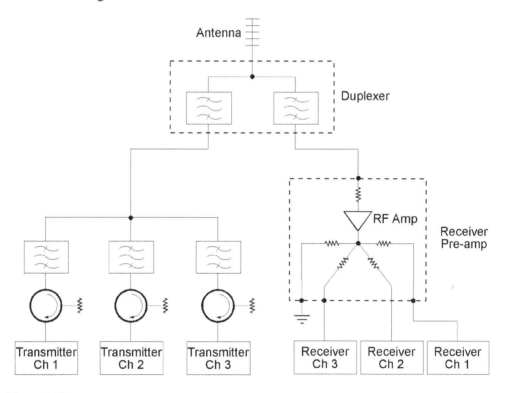

Figure 1.84
Typical equipment configuration for a large radio installation

1.16 Implementing a radio link

An important methodology must be followed when designing and implementing a radio link, if it is to work satisfactorily. It is relatively straightforward and will provide successful radio communication, if followed closely.

1.16.1 Path profile

The first requirement in establishing a successful radio link is to draw up a radio path profile. This is basically a cross-sectional drawing of the earth, for the radio propagation path, showing all terrain variations, obstructions, terrain type (water, land, trees, buildings, etc) and the masts on which the antennas are mounted. For distances less than a kilometer or two, profiles are not normally required, as the RTU can quite often be clearly seen from the master site (but all other calculations and choices described in the design methodology must be carried out).

Figure 1.85 illustrates a typical path profile.

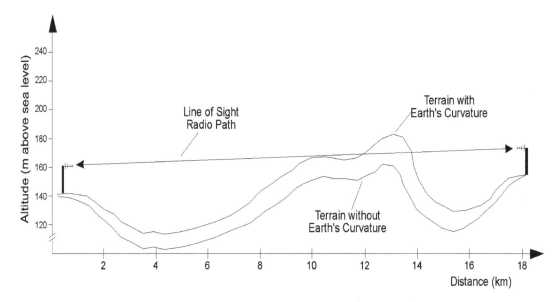

Figure 1.85
Typical path profile

The first step in this process is to obtain a contour map of the location. For most areas of developed countries and many areas in developing countries, these survey maps are readily available from government departments that oversee land administration and private companies that carry out surveys and publish their material. It is recommended that the map have a minimum of 20 m contour lines, with 2 m, 5 m or 10 m being preferred.

Locate the RTU and master site locations on the map and draw a ruled line between the two locations with a pencil. Then assuming that the master site is at distance 0 km, follow the line, noting the kilometer marks, and where a contour line occurs, and at that point note also the contour height.

The surface of the earth is of course not flat but curved. Therefore, to plot the points you obtained from the map directly would not be a true indication of the path. The formula below provides a height correction factor that can be applied to each point obtained from the map, to mark a true earth profile plot.

$$h = \frac{d_1 \times d_2}{12.75\,K}$$

Where

h = height correction factor that is added to the contour height (in meters)
d_1 = the distance from a contour point to one end of the path (in kilometers)

d_2 = the distance from the same contour point to the other end of the path (in km)
K = the equivalent earth radius factor

The equivalent earth radius factor K, is required to account for the fact that the radio wave is bent towards the earth because of atmospheric refraction. The amount of bending varies with changing atmospheric conditions and therefore the value of K varies to account for this. K factor is discussed in detail in section 2.14.1.

For the purposes of radio below 1 GHz, it is sufficient to assume that for greater than 90% of the time K will be equal to 4/3. To allow for periods where a changing K will increase signal attenuation, a good fade margin should be allowed for (refer to section 1.16.5).

The K factor allows the radio path to always be drawn in a straight line and adjusts the earth's contour height to account for the bending radio wave. Once the height has been calculated and added to the contour height, the path profile can be plotted.

From the plot, it can now be seen if there are any direct obstructions in the path. As a rule, the path should have good clearance over all obstructions. There is an area around the radio path that appears as a cone that should be kept as clearance for the radio path. This is referred to as the Fresnel Zone.

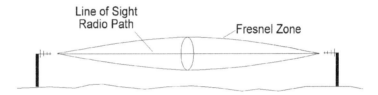

Figure 1.86
Fresnel zone clearance

Fresnel zone clearance is of more relevance to microwave path prediction than to radio path prediction, and will be discussed in greater detail in section 2.14.

The formula for the Fresnel zone clearance required is:

$$F = 0.55 \sqrt{\frac{d_1 d_2}{f(\text{MHz}) \times D}}$$

Where

F = Fresnel zone clearance in meters (i.e. radius of cone)
d_1 = distance from contour point to one end of path (in km)
d_2 = distance from contour point to other end of path (in km)
D = total length of path (in km)
f = frequency in MHz

If from the plot it appears that the radio path is going dangerously close to an obstruction, then it is worth doing a Fresnel zone calculation to check for sufficient clearance. Normally the mast heights are chosen to provide a clearance of 0.6 times ($\times$) the Fresnel zone radius. This figure of 0.6 is chosen because firstly, it gives sufficient radio path clearance and secondly, assists in preventing cancellation from reflections. At less than 0.6 F attenuation of the line of sight signal occurs. At 0.6 F, there is no attenuation of the line of sight signal and therefore, there is no gain achieved by the extra cost of providing higher masts.

Another important point to consider is that frequencies below 1 GHz have good diffraction properties. The lower the frequency the more diffraction that occurs. Therefore, for very long paths it is possible to operate the link with a certain amount of obstruction. It is important to calculate the amount of attenuation introduced by the diffraction, and hence determine the effect it has on the availability (i.e. fade margin) of the radio link.

As an example, Figure 1.87 shows a hill obstructing the radio path. Therefore, a calculation is required to be carried out to determine the attenuation due to diffraction, at this hill. This would be then added to the total path loss to determine if the link will still operate satisfactorily.

Mobile radios (moving RTUs) use a completely different set of criteria and formula. The greater majority of telemetry links are fixed and analysis of mobile radio is not provided in this book.

1.16.2 RF path loss calculations

The next step is to calculate the total attenuation of RF signal from the transmitter antenna to the receiver antenna. This includes:

- Free space attenuation
- Diffraction losses
- Rain attenuation
- Reflection losses

Free space attenuation is calculated using the formula given in section 1.4.5. Rain attenuation is negligible at frequencies below 1 GHz.

Reflection losses are difficult to determine. First, the strength of the reflected signal depends on the surface it is reflected off (e.g. water, rock, sand). Secondly, the reflected signal may arrive in phase, out of phase or at a phase angle in between. So reflected waves can be anything from totally catastrophic to enhancing the signal. Good engineering practice should always assume the worst case, which would be catastrophic failure. Therefore, when designing a link, a check is made for reflections and if they exist, measures should be taken to remove the problem.

This can be done by moving antennas or masts to different locations and heights or by placing a barrier in the path of the reflection to absorb it. For example, place the antenna behind a hill, house, billboard, etc.

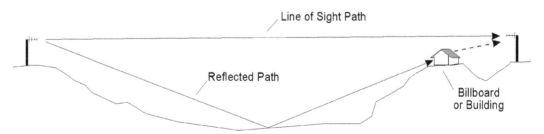

Figure 1.87
Removing potential reflections using barriers

Therefore the total RF loss is (a) + (b).

With reference to Figure 1.88, the total loss would be:

(a)
$$A = 32.5 + 20\,Log_{10}\,F + 20\,Log_{10}\,D = \text{Free space loss}$$
$$= 32.5 + 53.1 + 24.6$$
$$= 110.2\,\text{dB}$$

(b) Diffraction loss $= 23$ dB

$$(a) + (b) = 133.2\,\text{dB}$$

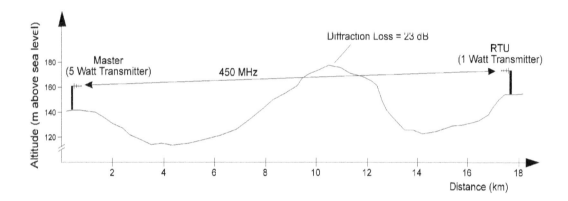

Figure 1.88
Example link

1.16.3   Transmitter power/receiver sensitivity

The next step is to determine the gain provided by transmitters. If in a link-configuration one transmitter operates with less power than the other does, the direction, with the least power transmitter, should be considered. Therefore, as ACA regulation requires that RTUs be allowed to transmit a maximum of 1 watt into the antenna, while master stations can transmit 5 watts into the antenna (sometimes higher), then the path direction from the RTU to the master should be considered.

The transmit power should be converted to a dBm figure. For an RTU this would be as follows:

$$Power = 10\,log\left(\frac{1}{10^{-3}}\right)\text{dBm}$$

$$Power = +30\,\text{dBm}$$

The next step is to determine the minimum RF level at the receiver input that will open the front end of the receiver (i.e. turn it on). This is referred to as the receiver threshold sensitivity level or sometimes as the squelch level. This figure can be obtained from the manufacturer's specification sheets.

For a radio operating at 450 MHz, this would be approximately −123 dBm. At this level, the signal is only just above noise-level and is not very intelligible. Therefore, as a general rule a figure slightly better than this is used as a receiver sensitivity level. A *de facto* standard is used where the RF signal is at its lowest and still intelligible. This level is referred to as the 12 dB SINAD level. This is discussed in the next section.

Again, this figure is obtained from a manufacturer's data sheets. For a typical 450 MHz radio, this level is approximately −117 dBm.

Using these figures, a simple calculation can be performed to determine the link's performance for the example link in Figure 1.88.

$TxPwr$ = Transmit power at RTU
$\qquad$ = +30 dBm
$Loss$ $\quad$ = RF path attenuation
$\qquad$ = 133.2 dB
$Rx\ Sen$ = Receiver sensitivity for 12 db SINAD
$\qquad$ = −117 dB
$\therefore Tx\ Pwr - Loss$ = Available Power at Receiver
$\qquad\qquad\qquad$ = +30 − 133.2 = −103.2 dBM

Since the receiver can accept an RF signal down to −117 dBm, the RF signal will be accepted by the receiver. In this case, we have 13.8 dBm of spare RF power.

1.16.4 Signal to noise ratio and SINAD

The most common measure of the effect of noise on the performance of a radio system is signal to noise ratio (SNR). This is a measure of the signal power level compared to the noise power level at a chosen point in a circuit. In reality, it is the signal plus noise compared to noise.

$$\therefore SNR = \frac{Psignal + Pnoise}{Pnoise}$$

SNR is often expressed in dBs. Therefore:

$$SNR = 10\ log_{10}\left(\frac{Psignal + Pnoise}{Pnoise}\right)$$

The real importance of this measurement is at the radio receiver. As the receive signal at the antenna increases, the noise level at the audio output of the receiver effectively decreases. Therefore, a measure of SNR, at the receiver audio output, for measured low level of RF input, to the receiver front end, is a good measure of the performance of a receiver.

The highest level of noise, at the audio output, will be when the RF input signal is at its lowest level. That is when the RF input signal just opens the receiver. At this point, the SNR will be between 3 and 6 dB.

A measurement has been developed as a *de facto* standard that measures the receiver input signal level for an SNR of 12 dB at the audio output. It is often referred to as the

12 dB SINAD level (signal to noise and distortion). The measurement is made with a device called a SINAD meter.

The SINAD measurement is carried out by feeding the input of the receiver with an RF signal that is modulated by a 1 kHz input audio signal. The 1 kHz signal will produce harmonics and unwanted distortion in the audio output. A SINAD meter is placed at the receiver audio output. The SINAD meter measures the power level of the 1 kHz, plus the noise and distortion. It then filters the 1 kHz signal and measures the broadband level of noise-power without the 1 kHz signal. It then divides the two measurements and provides a reading of SNR in dBs. The level at the receiver RF input is slowly increased until the SNR at the audio output is 12 dB. The RF input level is then noted and this is referred to as the 12 dB SINAD level. Some manufacturers refer to it as sensitivity at 12 dB SINAD or sometimes, just 'sensitivity'. The latter can be very deceptive and care must be taken.

The equipment configuration for measurement of SINAD is shown in Figure 1.89.

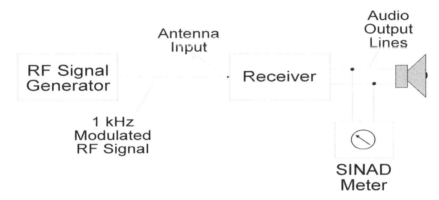

Figure 1.89
Equipment configuration for measuring SINAD

Some typical 12 dB SINAD sensitivity figures for modern radios in the VHF and UHF bands are 0.25 µV to 0.35 µV. The receiver opening sensitivity is normally 0.18 µV to 0.2 µV (i.e. the squelch level).

Another measurement also used to determine receiver performance is the '20 dB *Quieting*' measurement. This is not used as often as the 12 dB SINAD method.

Here, the receiver squelch is set so that it is just open. A level meter is connected to the speaker and the noise level measured. The audio volume control is set for a convenient level (0 dBm). A signal generator is then connected to the receiver input and an unmodulated RF signal is fed into the receiver, and increased slowly until the noise level at the audio output has dropped by 20 dB.

The disadvantage of this method is that it only measures the ability of the receiver to receive an un-modulated RF carrier signal. Poor design, component aging, or improper alignment, can provide a circuit response that admits the carrier signal perfectly but provides poor quality reception of modulated signals.

1.16.5   Fade margin

Radio is statistical by nature and it is therefore impossible to predict 100% how it is going to perform. For example, due to the degrading effects of reflections, multipathing, ducting, and RF interference, a link may lose or gain signal by up to 15 dB over short, or long periods, of time. It is because of this unpredictability that it is important to have a safety margin to allow for intermittent link degradation. This safety margin (or spare RF power) is generally referred to as the **fade margin**.

Idc recommend that it should be the intention to design most links to have a fade margin of approximately 30 dB. This means that if there was a 30 dB drop in RF signal level, then the RF signal at the receiver input would drop below the 12 dB SINAD sensitivity.

Therefore, in the example in section 1.16.3 there is insufficient fade margin. This is overcome by using high gain antennas.

For example, if we use a 13 dB gain Yagi at the RTU and a 6 dB gain omni-directional antenna at the master site, we add an extra 19 dB gain to our signal.

Therefore, the total fade margin in our example would become:

$$13.8 + 19 = 32.8\,dB$$

Finally, we must consider other losses introduced by cables, connectors, multicouplers, etc. In this example if we have 20 m of 3 dB/100 m loss cable, at each end, total connector losses of 0.5 dB at each end, and a multicoupler loss of 3 dB at the master site, then :

$$\begin{aligned}
Extra\ losses = Cables &: 2\,(0.2 \times 3\,dB) \\
&= 1.2\,dB \\
Connectors &: 2 \times 0.5\,dB \\
&= 1\,dB \\
Multicoupler &= 3\,dB \\
\therefore Total\ extra\ losses &= 3 + 1 + 1.2 \\
&= 5.2\,dB
\end{aligned}$$

Therefore the fade margin for the link is:

$$32.8 - 5.2 = 27.6\,dB$$

1.16.6 Summary of calculations

The following equation is a summary of the requirements for calculation of fade margin:

Fade margin = – (free space attenuation) – (diffraction losses) + (transmitter power) + (receiver sensitivity) + (antenna gain at master site) + (antenna gain at RTU) – (cable and connector loss at master) – (cable and connector loss at RTU) – (multicoupler filter or duplexer loss) + (receiver pre-amplifier gain).

1.16.7 Miscellaneous considerations

When implementing a radio link there are other important considerations. The following is a list of some of the main considerations.

a) It is important to obtain specification sheets from radio suppliers before purchasing any equipment and ensuring that all parameters meet your requirements.

b) The audio frequency output and input for a radio is normally a balanced 600 ohm connection. Depending on the equipment provided, it will accept levels, of between −30 dBm and +15 dB, and will output levels, from −15 dBm to +15 dBm. These are normally adjustable internally.

c) Most radios operate at +12 volt dc (or 13.8 V dc if it is floated across a battery). Depending on the RF output power level the current consumption may vary from 1 amp for 1 watt output to 10 amps for 50 watts output. This should be taken into consideration when sizing power supplies and batteries.

d) The size and weight of the radio equipment should be noted so that correct mechanical mounting and rack space can be provided. Multicouplers and duplexers can be very large and may require mounting on walls outside the radio equipment racks, or on separate racks. In order to minimize intermodulation, multicouplers and filters should be as close, as possible, to the transmitter and receiver.

e) It is beneficial to obtain 'mean time between failure' (MTBF) figures from manufacturers, because if the radio is to be placed in a remote location that is difficult to regularly acccss, it is vital that the radio is reliable.

f) If the telemetry communication protocol requires the radio to be switched on and off at regular intervals, it is best to avoid having relays in the inline RF circuit. Discrete transistor RF switching is preferred.

1.17 Types and brands of radio equipment

1.17.1 Types

For the purposes of implementing telemetry systems, there are four (main) classifications of radio equipment available.

1.17.1.1 Base station

This is a high quality radio designed mainly for use at master stations. They are generally designed and built to provide higher RF power for longer periods. They will have very stable oscillators, low noise receivers, be designed for high reliability and less interference, and have numerous other features, facilities and options. For example, they will normally have remote fault diagnostic facilities so that the status of the radio can be determined from a distant central location. They will also normally have options for duplication.

1.17.1.2 Mobile radios

This is the radio that is placed in a vehicle and used for mobile communications. It is quite common for these to be used (in a slightly modified configuration) as the radio source at the RTU. If they are being used in a simplex or half duplex mode, then minimal modification is required. If they are to be used in full duplex mode, then a duplexer must be added.

Mobile radios, because of their compactness, are inherently not as reliable, nor are their performance specifications as good as base stations. Since the RTU radio is not as vital as the master site radio, (the system will continue to operate if the RTU radio fails but not so when the master station radio does), normally, a mobile radio will suffice.

1.17.1.3 Mobile/base radio

The next type of radio is a cross between a mobile and a base station. They are normally the RF side of the mobile, fitted into the control electronics of a base station. They are often advertised as base stations at a cheaper price.

If a radio is required at the RTU, which is more reliable and has better performance than a mobile, then these radios can be a good option. Idc do not recommend their use for master site radios.

1.17.1.4 Radio modems

A common method of transmitting digital data across a radio link is to use a Radio Modem. This is basically a radio and a modem in one box, both specially designed for telemetry applications.

With some of these units, the radios have been specially designed for data over radio (and generally will not carry voice). They are normally of a high quality with good performance specifications and reliability. They will provide data rates of 4800 and 9600 bps in the 400 to 520 MHz band and rates of 9600 and 19 200 bps in the 860 to 950 MHz band.

As the radio modem comes prepackaged with a modem, they make system implementation significantly easier. The trade-off, of course, is price and some brands can be very expensive.

Another advantage of this type of unit is that because they have been designed to specially carry data over radio, they are normally capable of higher data rates than normal mobile or base station radios.

These units are not strictly digitally modulated radio units as the digital signal is applied to an analog radio or in some cases the digital signal is converted to an audio signal and then applied to analog radio. This aside, they are normally very robust and reliable units that provide performance that is more than adequate for most industrial telemetry applications. Some of them operate very sophisticated over-air protocols to ensure that the data that is received is 100% correct.

Pure digitally modulated radio is discussed further at the end of this chapter and in Chapter 2 on 'Microwave radio'

1.17.1.5 Spread spectrum radio

(Refer also to section 3.4.2.3)

Spread spectrum type radio communications is where data is transmitted in a pseudo random manner across a wide bandwidth. The transmitter is programmed with a code to which only the user knows the key. This code is then used to selectively transmit data across a wide bandwidth using short bursts of data. For example, the bandwidth across which the transmitter is transmitting may be 20 MHz wide, with burst transmissions of 20 kHz wide and 200 m/seconds in length. These pseudo random bursts of data across the bandwidth are just seen as noise to an observer. The user then programs the receiver with the same pseudo random code as the transmitter, so that when the transmitter is transmitting on a narrow 20 kHz band the receiver will be listening to that same band at the same time.

There are effectively two types of spread spectrum radio. The first technique is referred to as 'frequency hopping spread spectrum'. This operates as described above. The allocated bandwidth is broken down into a number of segments. The transmitter then uses a pseudo random code to move around the bandwidth segments, transmitting the data as it hops from one segment to the next.

The second technique is known as 'direct sequence spread spectrum'. Again, the allocated bandwidth is broken up into segments. However, this time data is sent across all of the segments in patterns that allow several or all segments to be used simultaneously. This technique allows higher data rates, but is more susceptible to noise and interference.

Spread spectrum technology is used in a number of different frequency bands for a number of different radio applications. Three sections of bandwidth have been allocated for free-to-air spread spectrum applications (no license is required for use). These are in the 900 MHz band (exact allocation varies from country to country), in the 2.400 to 2.4835 GHz band, in the 5.150 to 5.350 GHz band and in the 5.725 to 5.825 GHz band.

The main advantage of spread spectrum is that you can obtain very high data rates.

The disadvantage is that it basically becomes a '*free for all*' and as traffic begins to increase in built up areas then interference and error rates begin to increase, slowing data throughput. Arguably, it is also a poor use of frequency spectrum. Many channels can be placed into the same bandwidth using conventional techniques.

Spread spectrum radio is discussed further at the end of this chapter.

1.17.2   Brands

There are hundreds of manufacturers of radio equipment around the world. Some of the more common ones include:

- Motorola
- Simoco
- Kenwood
- Kyodo
- Tait
- Trio
- Radiolab
- Plessey
- GE
- Nokia
- ICOM
- Standard
- Maxon
- UNIDEN
- Midland
- Kachina
- Yaesu
- Nutel
- Barrett
- RF Systems
- Spectra

Note that this list is not all-inclusive.

1.18 Data transmission over analog radio

Telemetry by nature is data over radio technology. All telemetry systems support the transmission of digital information, from data acquisition centers to master locations and visa versa. With radio telemetry, the digital information is coded into a form that can be accepted by the radio for transmission. For normal audio frequency type radios, the digital information is, of course, coded into audio frequencies before transmission. This conversion of digital information into audio frequencies is carried out by a modem.

This section will discuss some of the problems encountered in converting digital information into audio frequencies and transmitting these over radio.

Starting to appear on the market today is digital radio. In this case, the digital information is NOT converted to audio frequencies but is applied directly to the carrier wave, for transmission. There are many advantages in using this technique but the technology is complex and expensive.

1.18.1 Modulation techniques

The most common modulation type used for transmission of digital information over analog radio is frequency shift keying (FSK). To a lesser extent differential phase shift keying (DPSK), multilevel phase shift keying (M-PSK) and amplitude shift keying (ASK) are also used. A combination of DPSK and ASK, known as quadrature amplitude modulation, is sometimes used, but not with great success at high speeds.

With FSK, data is sent by encoding a logic-1 with one tone, say 1200 hertz, and a logic-0 with a second tone, say 2200 hertz. Therefore, an FM radio transmitting data at 1200 baud will have a modulation index of:

$$\frac{1000}{1200} = 0.83 \text{ Hz/baud}$$

A second form of FSK is known as minimum shift keying (MSK). Here the frequency shift between the two logic states is reduced to half. Therefore, the modulation index becomes 0.415.

MSK is used with some systems requiring high data rates. MSK is only really suited to radio systems and is generally not provided in standard commercial modems.

Using FSK, DPSK and QAM, speeds of 300, 600 and 1200 baud are easily obtained over standard radio links. A speed of 2400 baud is possible with radios of good quality design, and with low RF and audio distortion.

Using DPSK and MQAM on analog radios is not very successful, because of their non-linear responses. Fast frequency shift keying (FFSK) and MSK are the two most common techniques used on standard analog radios. DPSK and MQAM are used on direct modulation digital radio (microwave and future mobile).

To obtain a speed of 4800 baud from a standard radio, modifications of the radio are normally required, to improve the linear performance of certain components, and to reduce the overall audio and RF distortion. A speed of 9600 baud, in a 12½ or 25 kHz bandwidth analog radio, is normally only obtained using a specially built radio with high quality linear components and digital software filtering.

There are a number of radio modems available in the market, which operate successfully at 9600 baud.

1.18.2   Transmission limitations

Most radios that are used for voice transmission operate with very high levels of distortion. Because of the logarithmic sensitivity of the human ear, it is very tolerant of distortion and noise, and this is generally accepted without questions. For example, it is not uncommon for an audio output at a receiver to exhibit 5% total harmonic distortion (THD).

Although this may be acceptable for normal voice applications, it can have disastrous effects on the transmission of digital information.

The following is a description of the main causes of problems in transmission of digital information over analog radio. Overcoming these problems is achieved quite simply by good radio design.

1.18.2.1   Frequency response

Most standard radios have non-uniform audio frequency versus amplitude responses. This is illustrated in Figure 1.90.

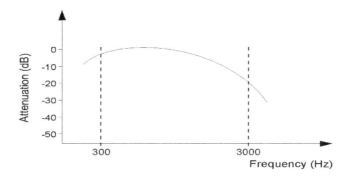

Figure 1.90
Audio frequency response of radio

This may be caused by high frequency attenuation in the audio amplifier or by a very narrow IF filter. This in turn will distort the data output.

If the IF filter has a rippled response in its pass band, the output audio frequency response is further deteriorated. Therefore, a well designed IF filter with a flat pass band, and a low distortion audio amplifier, is required for data-over-radio applications.

It should also be noted that all pre-emphasis and de-emphasis circuits should be removed from the radios for data transmission.

1.18.2.2   Phase shift distortion

Phase shift distortion, also known as 'envelope delay distortion', is where the propagation time of different frequencies through an audio communications link varies. For example, a 1 kHz frequency may take 0.5 ms to propagate through a radio link while a 2 kHz may take 1.5 ms to propagate through the same link. This plays havoc with a digital signal waveform, as all associated harmonics become distorted. Figure 1.91 illustrates the effects of non-linear phase responses in radio. The term 'envelope' refers to the response of the channel over the complete frequency range.

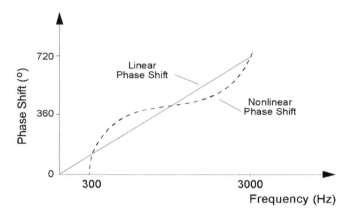

Figure 1.91(a)
Typical phase response of a communication channel

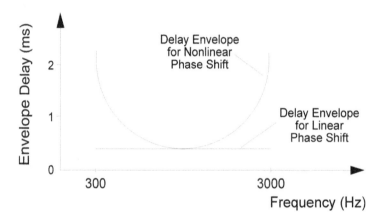

Figure 1.91(b)
Propagation delay of audio frequencies

1.18.2.3 Amplitude response distortion

Amplitude response distortion occurs where there is a non-linear amplitude response across a link. This is illustrated in Figure 1.92.

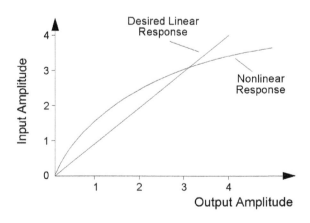

Figure 1.92
Example of non-linear amplitude response

This type of non-linear amplitude response can cause clipping, which in turn produces spurious harmonics that will devastate a data communications circuit. Non-linear amplitude responses are particularly damaging to modulation schemes such as QAM that use quantized changes in amplitude, for coding information.

1.18.2.4 Frequency translation error

With an FM radio, where the carrier is modulated with distinct tones of digital information, the receiver will receive those exact modulated tones. A problem occurs when the modulator and the demodulator carrier frequencies are slightly mistuned, then the demodulated output signal will be slightly off tune or distorted. For AM radio, this problem is even more critical.

1.18.2.5 Jitter

There are two main forms of jitter:

- Phase jitter
- Amplitude jitter

Jitter is the recurrence of very short sharp changes in phase or amplitude. Jitter can be introduced at any point of a communications link. It is generally caused by noise getting into a modulator or demodulator, and can cause very serious distortion of a signal.

1.19 Regulatory licensing requirements for radio frequencies

For the relevant regulatory requirements and standards associated with the use of the frequency spectrum in your country check with the appropriate authorities.

1.20 Duplication and self-testing

1.20.1 Duplication

If a telemetry system is an important element of an industrial process or application, and it is considered vital to the integrity and/or safety of the industrial process that the radio link does not fail, then it is possible to **duplicate** a part or all of the system. Duplication, in radio terminology, implies that there are two identical components in a system and if one fails, the other immediately comes into operation to replace it.

With a radio system, it is possible to have various levels of duplication. For example, the transmitters are usually the most common component to fail and therefore they are usually, the first part of a radio system to be duplicated. The two most common configurations for duplicated radio systems are, firstly, complete duplication of all equipment, and secondly, duplication of the transmitter/receiver units.

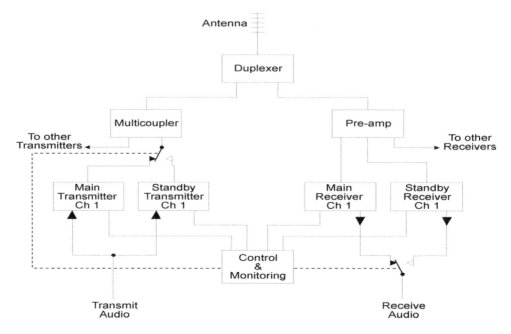

Figure 1.93(a)
System with duplication of transmitter and receiver

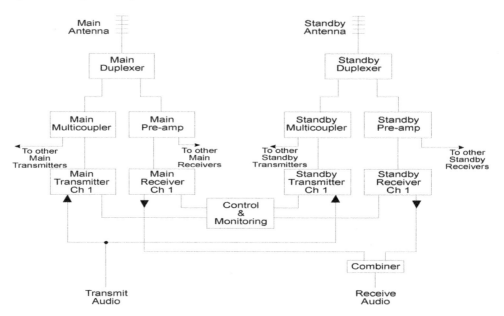

Figure 1.93(b)
Fully duplicated radio system

Figure 1.93(a), shows a central control and monitoring unit that monitors the status of the main and standby, transmitters and receivers, and if a unit fails, it activates a relay to change to the standby units.

Figure 1.93(b) shows a system where all components are duplicated.

The control and monitoring unit, monitors all the equipment and chooses which transmitter to use, and which audio to receive.

1.20.2   Standby transmitters

There are three types of standby transmitters. These are:

Hot

This is one where the standby transmitter is operating continuously into a dummy load. When the main transmitter fails, the output of the standby is switched over to the antenna. In this mode, it is possible to continuously monitor the state of both transmitters.

Warm

This is one where the standby transmitter is not keyed up at the final power amplifier stage so that it can remain connected to the antenna, but all other components of the transmitter are active (e.g. oscillators). When the main fails, the power amplifier is simply keyed up.

Cold Standby

Here the standby unit has no dc power applied to it until the main transmitter fails.
Standby modes are discussed further in Chapter 2.

1.20.3   Self-testing and remote diagnostics

Most modern radio basestations have the facility to automatically test certain operating parameters at regular intervals or to carry out manual testing on an as required basis. Some parameters that can be tested and monitored include Tx output RF power, Rx input receive level, VSWR, power supply current and voltage, modulation level and bit error rate. If the unit detects a fault, then alarms are initiated locally and sent to a central site monitoring station. The basestation unit will then automatically change from main to standby (assuming it is a duplicated site). The status information will be relayed back to the central site monitoring facility, either via landline, microwave or over the radio itself. A central operator is then able to monitor the status, and manage the operation, of a number of remote base stations. If required, the operator can also send control information to the sites to carry out changeover of duplicated basestations, switch from voting to TTR operation, disable certain elements of the system or one of many other controlled activities. This is generally referred to as a remote diagnostics facility.

1.21   Miscellaneous terminology

1.21.1   VSWR and return loss

Two concepts often referred to in radio are return loss and voltage standing wave ratio (VSWR), both of which are closely related. To explain these terms, consider a normal coaxial cable with a radio frequency wave traveling down its length. When the wave reaches the end of the cable, it can be, totally absorbed, totally reflected or partially absorbed/reflected, at some state in between.

The coaxial cable itself will have its own natural impedance to this wave, referred to as the cable characteristic impedance Z_o. The characteristic impedance will vary a small amount with frequency.

When a load is placed on the end of the cable, with impedance exactly equal to the impedance of the coaxial cable, all the wave energy will be absorbed by the load. If there is a mismatch in impedance at the load connection point, then some of the energy is reflected back along the cable in the opposite direction. The amount of reflected energy depends on the mismatch in loads.

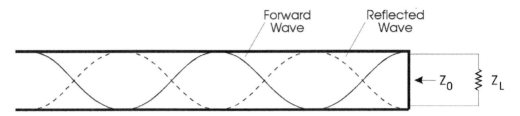

Figure 1.94
Wave energy in a cable

If the cable is monitored at a single point along its length, when it is connected with a perfectly matched load, it will be noted at this point that the wave is continuously rising and falling in magnitude, as it travels along the cable. Therefore, the average voltage at any time on the cable is zero. However, when there is a mismatch in loads and there are reflected waves in the opposite direction, then the changing waves continually add and subtract each other at every point along the cable. This results in standing waves.

Most reflection of energy occurs when the load is short circuit (0 ohms) or open circuit (¥ ohms). Figure 1.95 illustrates how standing waves are formed from forward and reflected waves. In simple terms, the standing wave ratio is the larger of the two.

$$SWR = \frac{Z_L}{Z_o} \text{ OR } \frac{Z_o}{Z_L}$$

Where:

Z_L = Load impedance
Z_o = Cable characteristic impedance

By measuring the value of the standing wave voltage at the node (where the resultant voltage swings from positive to negative) compared to the voltage at the anti-node (where the resultant voltage is always zero) it is possible to determine the amount of energy that has been reflected. This is referred to as the voltage standing wave ratio (VSWR).

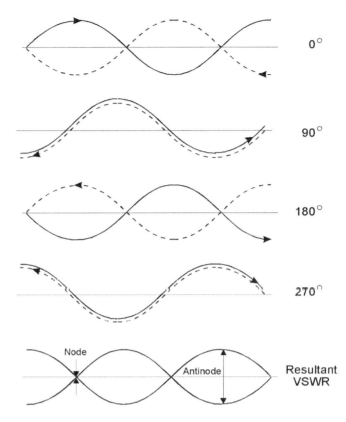

0°

90°

180°

270°

Resultant
VSWR

Figure 1.95
Formation of a standing wave from forward and reflected waves

Consider the case where a coaxial cable is connected to an antenna. A perfectly matched antenna would absorb the entire wave and radiate the energy into free space, but as a rule, antennas and coaxial cables are never a perfect match and some energy is reflected back down the cable.

By using an RF power meter, and measuring the forward power and the reflected power, it is possible to determine the VSWR.

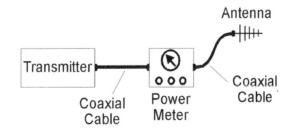

Figure 1.96
Connection of a RF power meter for measurement of forward and reflected power

The equation used to determine VSWR is:

$$VSWR = \frac{1 + \sqrt{\dfrac{Reflected\ power}{Forward\ power}}}{1 - \sqrt{\dfrac{Reflected\ power}{Forward\ power}}}$$

Another term often used in radio is return loss. This is a comparison of the forward energy to the reflected energy. The equation for return loss is as follows:

$$Return\ loss = \sqrt{\frac{Reflected\ power}{Forward\ power}}$$

$$= \frac{V_{Ref}}{V_{For}} \quad (When\ Z_{Ref} = Z_{For}\ )$$

Return loss is usually expressed in dB; therefore, the following two equations are also used:

$$VSWR = \frac{1 + Return\ loss}{1 - Return\ loss}$$

And

$$Return\ loss\ (dB) = 20\ Log_{10}\left(\frac{VSWR + 1}{VSWR - 1}\right)$$

Figure 1.97 illustrates the effect that VSWR has on forward and reflected power.

It should be the aim during radio systems design and implementation, to achieve a VSWR as close to 1.00 as possible, so that the radio system is operating at maximum efficiency. In real terms, reflected power is wasted power and detracts severely from the reliability and integrity of a radio link.

The absolute maximum VSWR that should be tolerated in any radio system is 3.0. A well-engineered radio system will have a VSWR of less than 1.5.

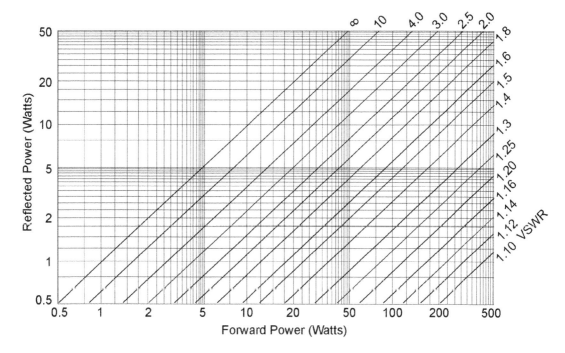

Figure 1.97
Graph of forward power versus reflected power for different VSWR values

The following table compares VSWR, with the percentage of reflected power and the comparative return loss that this reflected power represents.

VSWR	% of Power reflected	Return loss (dB)
1.01		46.1
1.02	0.01	40.1
1.03	0.02	36.6
1.04	0.04	34.2
1.05	0.06	32.3
1.06	0.08	30.7
1.07	0.11	29.4
1.08	0.15	28.3
1.09	0.19	27.3
1.10	0.23	26.4
1.11	0.27	25.6
1.12	0.32	24.9
1.13	0.37	24.3
1.14	0.43	23.7
1.15	0.49	23.1
1.16	0.55	22.6
1.17	0.61	22.1
1.18	0.68	21.7
1.19	0.75	21.2
1.20	0.83	20.8
1.30	1.70	17.7
1.40	2.78	15.6
1.50	4.00	14.0
1.60	5.33	12.7
1.70	6.72	11.7
1.80	8.16	10.9
1.90	9.63	10.2
2.00	11.10	9.5
2.20	14.1	8.5
2.40	17.0	7.7
2.60	19.8	7.0
2.80	22.4	6.5
3.00	25.0	6.0
3.50	30.9	5.1
4.00	36.0	4.4
5.00	44.4	3.5
6.00	51.0	2.9
7.00	56.2	2.5
8.00	60.5	2.2
10.00	66.9	1.7
20.00	81.9	0.9
50.00	92.3	0.3

Table 1.6

1.21.2   Desensitization and blocking

1.21.2.1   Desensitization

At a site where there are a number of other radios all operating from the same mast, it is possible for RF energy from other transmitters to cause interference to a co-located receiver's front end. This RF energy may cause the RF sensitivity level, at which the radio receiver normally opens, to increase. This effect is known as receiver desensitization.

For example, under normal conditions a receiver may open at 0.2 μV. However, when it is desensitized from other transmitters the receiver may take 0.4 μV to open. This will severely deteriorate the performance of a telemetry link, for remote RTUs that are being received at low levels from the master site.

Desensitization problems are dramatically reduced by the use of multicouplers, filters, and duplexers.

The common method used to test for desensitization is to switch on all the transmitters at a site, measure the level required to open the required receiver with an RF test set, then switch off the transmitters and measure the receiver opening sensitivity again. A well-engineered system will have little or no difference in levels.

1.21.2.2   Blocking

Blocking is the phenomena where a very strong radio signal, at a frequency that is different to (but normally in the same band as) the required receiver, saturates the receiver so that its phased-locked loop locks onto the strong interfering signal. This effect is noted when the car radio is passing a broadcasting radio station transmitter and the car radio loses the station it was tuned to and locks on to the station it is passing.

In a telemetry application, this effect may occur when a mobile radio, in a vehicle parked next to an RTU, is keyed up and the RF signal saturates the RTU receiver.

1.21.3   Continuous tone coded sub-audible squelch (CTCSS)

CTCSS is a method used to provide protection to receivers, from transmitters on the same frequency that are not part of the telemetry system. These rogue transmitters may be from vehicle mobile radios, operating outside their normal location, or transmitters at a significant distance that are interfering because of ducted signals.

With the CTCSS system, a very low frequency sub-audible tone (between 50 and 250 Hz) is transmitted with the radio data signal. The receiver has a very selective filter tuned exactly to this frequency. If the correct tone is in the received RF signal, the receiver will open and the information signal will be heard. If the tone does not exist then the receiver will stay closed. Therefore, transmitters that do not have the correct CTCSS tone and do not belong to the system will not open the receiver and cause interference.

Of course, this system of protection fails when the receiver is already open and receiving signals from transmitters that belong to the system, and a rogue transmitter keys up over the top of them.

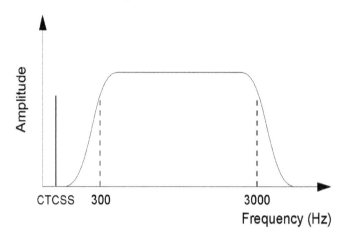

Figure 1.98
Illustration of the position of a CTCSS frequency in the audio band

There are a number of CTCSS tone sets available. One of the more common sets is published by the Electronic Industries Association (EIA). Most of the frequency sets have approximately 50 frequencies in them. Therefore, chances of the rogue transmitters using the same CTCSS frequencies are slim. There are no regulations specifying which frequencies are to be used.

A typical tone set is listed in Table 1.7 below.

1	67.0 Hz	11	94.8 Hz	21	131.8 Hz	31	171.3 Hz	41	203.5 Hz
2	69.4 Hz	12	97.4 Hz	22	136.5 Hz	32	173.8 Hz	42	206.5 Hz
3	71.9 Hz	13	100.0 Hz	23	141.3 Hz	33	177.3 Hz	43	210.7 Hz
4	74.4 Hz	14	103.5 Hz	24	146.2 Hz	34	179.9 Hz	44	218.1 Hz
5	77.0 Hz	15	107.2 Hz	25	151.4 Hz	35	183.5 Hz	45	225.7 Hz
6	79.7 Hz	16	110.9 Hz	26	156.7 Hz	36	186.2 Hz	46	229.1 Hz
7	82.5 Hz	17	114.8 Hz	27	159.8 Hz	37	189.9 Hz	47	233.6 Hz
8	85.4 Hz	18	118.8 Hz	28	162.5 Hz	38	192.8 Hz	48	241.8 Hz
9	88.5 Hz	19	123.0 Hz	29	165.5 Hz	39	196.6 Hz	49	250.3 Hz
10	91.5 Hz	20	127.3 Hz	30	167.9 Hz	40	199.5 Hz	50	254.1 Hz

Table 1.7
CTCSS tone set

1.21.4   Selective calling

Selective calling, or Selcal as it is often referred to, is a system of enabling each individual radio in a system of radios operating on the same frequency, to be called and communicated with, without the other radios hearing the communication. Each radio is given a separate five-digit code. Each digit is transmitted in the form of audio tones (often a DTMF tone pattern is used).

This enables a master site to access a single RTU without initiating communications with other RTUs. All sites hear the transmitted five-tone sequence, but only the RTU, which is registered with the transmitted code, will respond.

Most modern mobile and base station radios are provided with Selcal facilities. They are part of programmable facilities provided in an internal EPROM.

1.22 Additional features and facilities for analog radio

The standards of radio technology used in industry were, for many years, stagnant, as operators became conservative in their expectations. Over the past 10 years there have been significant advances in radio technology. A prudent selection from the vast array of features now available can greatly increase work efficiency and system integrity.

Some of these features include:

- Radio/telephone interfacing
  Accessing the radio system from the telephone network for voice and data transmission (this includes ISDN interface options).
- Selective calling (SELCAL)
  The use of individual codes to access single radios or groups of radios. (This enables private data flow between two or more radios. *Refer to* section 1.21.)
- CTCSS
  The use of sub-audible tone to help prevent unwanted interference from outside operators. (*Refer to* section 1.2.1.)
- ANI
  Automatic number identification (ANI). Every radio in the system is programmed with an individual Selcal number. These are used as identification codes. The identification codes can then be correlated to a vehicle or location where the radio is installed. The Selcal number is always transmitted as soon as the radio is keyed up. Therefore, if a particular radio or radio user is causing problems on the system and the source cannot be identified, decoding of the Selcal number using ANI will reveal who/where a problematic transmission is coming from.
- Scanning
  The radio unit is programmed to continually scan over a number of receive frequencies and to lock on to the first frequency that it comes across, on air. It does this in a sequential manner, locking on to the first frequency that it comes across, and not discriminating between frequencies if two frequencies become active at the same time. Some radios have the ability to setup a priority scan. Here, if the radio is required to scan over a large number of channels then the frequencies of certain high priority channels are checked more than once, during the complete scan cycle.
- Voting
  This system of frequency selection is similar to scanning. It is used where there are a number of frequencies being transmitted for one channel, generally to provide continuous coverage over a large geographical area. The unit is programmed to scan over the particular number of receive frequencies of the one channel. It measures and makes a note of the signal strengths of each frequency. It then selects the frequency with the strongest signal strength, which is then used by the radio unit for reception on the channel. It then also selects the transmit frequency, which is the corresponding receive frequency

at the repeater of the chosen channel, for transmission over the channel by the radio unit.

- PC programmable
  Most modern radios can be programmed for frequency and the many other features mentioned in this section, using a PC that is plugged into a small port in the front of the radio. The software for carrying this out is provided by the radio supplier.

- Data port
  Some brands of radios now provide a RS-232 port for attachment of a modem or fax. Data rates are generally slow, in the order of 1200 to 2400 Baud. Some higher quality radios that have been specifically manufactured for data will allow data rates of 9600 and 19 200 Baud.

- DTMF keypad access
  As an option, many manufacturers can fit 'dual tone multiple frequency' (DTMF) keypads to the front of the handheld or mobile. This can be used for Selcal purposes as discussed above or for access to a telephone network.

- Encryption
  Allows a high level of security for radio communications, by encoding voice and data before transmission to air, and decoding the received signal.

- Improved receive sensitivity
  Allowing lower signal strengths to be received with less noise. (This provides improved coverage area.)

- Improved transmitters
  Provide a more linear and less distorted RF signal transmission, which improves the quality of the transmitted signal allowing higher data rates.

- Synthesized radios
  Allows the radio to operate with more stability and provides greater frequency agility, over a greater range of frequencies, and more flexible selection from a larger number of channels.

- More efficient operation
  Uses less supply current for the inactive state and for higher transmit powers.

- More compact radio units
  Allowing installation in more convenient and less spacious areas.

- Remote diagnostics of TTR
  Allowing a distant terminal to access base radio stations, to check on their current operational status and to diagnose faulty equipment.

Some of the specific features of a base station radio include:

- Higher transmit powers (up to 100 watts)
- Longer continuous transmit periods allowed
- A high degree of frequency stability
- Low noise receivers
- High MTBF figures
- Low levels of spurious emission
- Low levels of intermodulation
- More linear phase, frequency and amplitude responses than mobiles and handheld
- Frequency synthesized

- Up to 100 channels selection
- Remote diagnostics and configuration via RF
- Remote diagnostics and configuration via landline
- Full duplex operation (dual synthesizers)
- Facilities for data transmission
- Selcal, CTCSS, ANI, scanning, voting
- Duplication options and automatic changeover on failure
- Automatic self-testing of transmit power, VSWR, modulation level, receiver sensitivity, supply voltages, operating temperature, TTR loopback, BERs

1.22.1 Trunking radio

A trunked radio system is one in which two or more radio channels are time shared by a larger number of user groups on the system. The trunked radio system uses an access channel (referred to as the control channel) through which a user group gains access to a pool of available channels and seizes one for their own use. It is very similar in concept to a modern day switched telephone network. As in a telephone network, the radio user must dial or otherwise generate the *address* of the user group for which the call is intended. The address sent to the central base station/switching equipment is translated to the user group for which the call is intended. The equipment automatically makes the connection and the call proceeds as in standard radio systems, until users have finished the conversation and the call is terminated. The user groups hear only the calls intended for them because of the selective calling built into the system.

The system is designed and dimensioned in a similar manner and using similar traffic theory that is used to design a telephone system.

The trunking system is described in a '1 + n' notation. The '1' represents the control channel, and the n the number of voice channels available to users.

In some trunked systems, the control channel can be used as a voice channel when all other channels are in use. This is referred to as a non-dedicated system. This allows more efficient use of the equipment and frequencies available.

Several trunking base stations can be linked together to form a network of master sites providing common facilities over a large area.

Some of the extra facilities available on a trunking system include:

- Easy system expansion
- Multiple PSTN connections
- Priority calling
- Multiple site networking and dynamic group allocations
- Individual or group calling
- Network management systems
- Remote fault diagnostics
- Automatic call queuing
- Traffic statistics
- Flexibility to change network and operational structures

There are a number of system standards used around the world. Several have been developed by individual companies and are proprietary. An international standard has been developed which is becoming more popular worldwide. This is known as the MPT 1327 signaling standard and the MPT 1343 performance standard.

Each system has its benefits and limitations and suitability depends more on the user's requirements.

1.23 Digital modulation radio

Over the last ten years, there has been a gradual evolution of radio communications towards digital modulation techniques. This started with point-to-point high capacity microwave radio communications, operating in the frequency bands above 1 GHz. Then point-to-multipoint microwave systems were developed. (*Refer to* Chapter 2.) The first half of the 1990s saw the development of digital cellular technology. This technology has now been successfully implemented in standard systems around the world for cellular mobile telephone communications.

1.23.1 Digital private mobile radio

Private mobile radio (PMR) technology has been a lot slower in the uptake of digital radio technology. Although several vendors have advertised products that are supposedly digital in nature, until very recently there has been little acceptance of these products. In addition, during the early part of the 1990s a number of proprietary standards and products were developed that were not well accepted. There are several standards currently under development, which are being proposed as international standards.

The main standards that are in use worldwide are as follows:

APCO-25	This standard has been pushed by North America but has had little uptake outside of the United States.
Tetra	This standard has been developed and pushed by Europeans. After many years of deliberation, there are finally a number of large Tetra systems about to be installed throughout Europe and other continents. This is the most likely standard to get widespread support.
Tetrapol	This standard was also developed in Europe and has a number of systems installed and widespread support from many manufacturers. The Tetra and Tetrapol standards are fundamentally different technologies. The standard is significantly less complicated than Tetra and generally less expensive to install. The down side is that it has limited features compared to Tetra.
EDACS	This proprietary standard was developed by Ericsson and, although there are not large numbers of systems installed, it has had widespread acceptance. It is generally considered a good system with many useful features. There has been some reluctance to install the system since it is proprietary and not an international accepted standard
Smarzone	This proprietary standard was developed by Motorola and there are many systems installed around the world. This also is generally accepted as a good quality system. Due to the modulation techniques used, there is some debate as to whether this is a true digital system or just a sophisticated analog trunking system.

The APCO 25 standard is coming out of America. This is being developed by a very influential public safety lobby group referred to as APCO. They officially do not release standards, but refer to as recommendations. Their first well known radio recommendation in the area of public safety was APCO–16. This was a user recommendation for operational procedures of trunked radio systems by public health and safety groups such as the police, fire brigades, ambulance, and strategic armed forces. This standard contained very little technical detail.

Around the late 1980s, APCO started work on the APCO 25 technical and operational standard for a digital modulation radio system for both trunked and single channel PMR. This standard was ratified in August of 1995. Manufacturers now provide a range of APCO 25 products.

The system uses digital signal processing to simulate the human voice and then only transmits the bits that represent a change in the human voice pattern, rather than the complete data description of the human voice (using straight eight bit resolution PCM for example). The software and microchips that carry out the processing of the human voice are referred to as *vocoders*. APCO-25 transmits over 12½ kHz bandwidth channels using a 9.6 kBit/s data rate (that is 0.77 bits per hertz). The modulation technique used here is referred to as C4 FM, which is a type of continuously shifted quadrature phase shift keying (QPSK).

APCO-25 was designed fundamentally to service the public safety service industry. It therefore operates like a standard two way radio system in nature. This is essential for quick radio access time required in this industry. Other industries that would be well served by radio equipment using this standard are the mining and manufacturing industries and to some extent, the utilities.

Data transfer capabilities are also included in the standard. The standard also includes many strict requirements for intrinsic safety and weather protection. It is also of interest to note that the digital signal processing standard recommended for use in the vocoder is designed to provide the least amount of distortion to the human voice when there is a high level of background noise such as helicopters, police and fire engine sirens, gun shots, etc.

The main standard coming out of Europe is referred to as the TETRA PMR standard. The TETRA standard was developed by the European Telecommunications Standards Institute (ETSI). The standard has finally been accepted by most European countries, although it has been through many years of debate and changes.

TETRA is designed more for the European PMR market where commercial applications are considered more important. The standard is designed to service many users who have commercial voice requirements and a heavy data transfer requirement.

The standard is based on time division multiplexing (TDM) technology, where four time slots are allocated to one 25 kHz bandwidth channel. The data transmission rate is 36 kbps over the allocated 25 kHz bandwidth channel (that is 1.44 bits per hertz). Here also digital signal processing is used with vocoders but this time the emphasis is on maximizing data throughput.

As this standard has specifically been developed for commercial applications, the transport industry, taxis, couriers, etc would best be served by this type of radio technology. A number of service utilities (electricity, water, municipalities, etc) would also be well suited. A number of country wide systems are to be installed throughout Europe and should be operational by 2004. A number of countries (for example Norway) currently have moderately sized systems up and operating successfully.

The Tetralpol standard was developed by the French and registered in 1994 as an open standard. It is based on the GSM standards, but is not directly accessible using GSM equipment. The standard uses Frequency Division Multiplexing (FDM) and therefore has very little in common with Tetra. It uses 12.5 kHz channel spacing and Gaussian minimum shift keying (GMSK) as its modulation technique.

The major benefits with using Tetrapol are that it is a simpler technology than Tetra, there is a lot of experience with manufacturers in producing GSM technology and that it is generally cheaper to implement than Tetra.

1.24 Digital radio specifically for telemetry applications

A number of manufacturers produce digital radios that are specifically for telemetry applications. They are designed to operate in the licensed bands to allow secure guaranteed throughput over long distances. Two of the most well known manufacturers of these radios are Trio Communications and Microwave Data Systems.

The philosophy behind the design of these radios is that they are 'work horses' for operation in remote or hostile environments where data integrity is vital. They are normally built as very rugged units, with sophisticated over-air protocols to ensure that all data received is 100% correct.

Some of the features of these types of radios include:

- Operate in the 400–520 MHz or 900 MHz telemetry bands
- Channel spacing of 12.5 kHz or 25 kHz
- Data speeds of 4800, 9600, or 19 200 bps
- Internal duplexers to allow full or half duplex operation
- Often use modulation techniques such as GMSK or QAM
- Options for fully duplicated internal components – power amplifiers, exciters, receivers, modems, CPU boards
- Automatic internal changeover of duplicated components upon failure of any component
- Automatic validation of changeover and sending of alarms to the remote monitoring facility
- Inbuilt RF lightning protection
- Front panel status displays controls for local control and ease of maintenance
- All parameters of the units will be fully programmable
- Local and remote monitoring, diagnostics, and control of the unit
- Provided with Windows based software packages for remote monitoring and configuration
- Purpose designed to minimize common components which could cause total failure conditions
- Hybrid combining networks rather than relays are used to ensure maximum reliability of radio frequency changeover circuits to ensure 'soft fail' operation.
- Use C/DSMA collision avoidance system so that multiple RTU radios can access the one base station
- Use sophisticated over-air protocols to carry out error detection and error correction to guarantee that only data that is 100% correct is forwarded on from the receiving radio
- Designed for 100% transmit duty cycle
- Operating temperature range of −10°C to +65°C or greater
- Operating humidity range of 0 to 95% or greater
- RF output power of 5 watts at the antenna port for full duplex continuous operation.

- Automatic RF output power reduction if ambient temperature goes over a defined level
- Often have two serial data ports for connection to different hosts.
- Often support serial interface protocols such PPP and SLIP.
- Often support industrial protocols such as Modbus
- Bit error rates of better than 10^{-6} at -108 dBm

1.25   Digital wireless communications

Since the mid 1990s, there has been an ever increasing interest in the development of what is often referred to as wireless communications. This generally refers to two areas of radio communications. The first is the area of licensed public mobile telephones that are operated by the larger telecommunications companies. Current second generation standards such as GSM fall into this category. The second area is that of free-to-air standards that are used for wireless networking. This is most commonly used for mobile computing and as a replacement for LAN cabling.

In the first category, there has been a lot of development towards 'third generation' cellular mobile telephones. These systems are designed to enable the user to have high data rate access to the Internet or other related data networks.

The full development and ratification of these standards has been slow to accomplish due to industry reservations as to the expected uptake of the services from the general public. At the time of writing this book, some telecommunications companies had begun the roll out of the first phase of their third generation networks. There is still some debate as to which standards will become the accepted standards worldwide.

Available now on the existing GSM networks is the general packet radio service (GPRS). This can provide up to 21.4 kbps. GPRS is used for some simple telemetry applications. For applications that do not require real time data flows and dial-up access is acceptable, then this service can work well. Latter model GSM phones or PCM CIA cards for computers have inbuilt GPRS modems. Otherwise, a GPRS modem can be purchased and attached to a standard GSM phone.

For the second category, there have been wireless LAN products around since the early 1990s. These have been successfully providing data speeds of 1 to 2 Mbps for LAN extension requirements.

Most of these products operate using some form of spread spectrum technology in the unlicensed 2.4 GHz band. Some are based on international standards while others are based on proprietary standards. Due to their use in the unlicensed band, their RF output powers are limited, which in turn limits their range of operation.

The main problem with using these systems is that number of users now operating in this band has become significant, particularly in built up urban areas. This is causing significant interference and considerably reducing the data throughput for most users. With the anticipated enormous growth in the area, it is expected that the problem can only get much worse.

A number of wireless LAN standards are now beginning to take precedence in this market. These are Bluetooth, providing data speed of 1 and 2 Mbps, 803.11b providing data speeds of 11 Mbps and 803.11a providing data speeds of 54 Mbps. Bluetooth and 803.11b will be operating in the 2.4 GHz band while 803.11a will be operating in the 5 GHz band.

The following table provides a brief overview of the various standards and how they compare with each other.

Technology	Geography	Date introduced	Frequency band	Multiple Access Technology	Modulation	Data rate
GSM (Global System for Mobile)	Mainly Europe & Asia, some Latin and North America	1992	450–467 MHz 479–496 MHz 747–792 MHz 824–894 MHz 876–960 MHz 1710–1880 MHz 1850–1990 MHz	TDMA	GMSK	14.4 kbps
GPRS (General Packet Radio Service)	An update of GSM	2001	As above	TDMA	GMSK	21.4 kbps
HSCSD (High Speed Circuit Switched Data)	An update of GSM	2000	As above	TDMA	GMSK	14.4 kbps
EDGE (Enhanced Data Rates for GSM Evolution)	An update of GSM	2001 – 2	As above	TDMA	8 PSK	384 kbps
TIA/EIA-136 (North American Digital Cellular)	North America and some Latin America	1996	824–894 MHz 1850–1990 MHz	TDMA	DQPSK	13 kbps
TIA/EIA-95-A (CdmaOne System)	North America, Korea and some Asian countries	1995 – 1997	824–894 MHz 832–925 MHz 1850–1990 MHz 1750–1870 MHz	CDMA	QPSK down OQPSK up	14.4 kbps
TIA/EIA-95-B (CdmaOne System)	North America, Korea and some Asian countries	1995 – 1997	824–894 MHz 832–925 MHz 1850–1990 MHz 1750–1870 MHz	CDMA	QPSK down OQPSK up	115 kbps
CDMA2000 (1xRTT) (1xRadio Telephone Technology)	Korea, USA and Japan	2001	411–493 MHz 1920–2170 MHz	CDMA	QPSK down HPSK up	307.2 kbps
1xEV-DO (1xEvolution Data Only)	Korea, USA and Japan	2001 – 2002	411–493 MHz 1920–2170 MHz	CDMA	QPSK/8PSK/ 16QAM down HPSK up	2.5 Mbps down 307 kbps up
W-CDMA (FDD) (Wideband CDMA Frequency Division Duplex)	Korea, Japan, Europe, USA and Asian Countries	2001	1920–2170 MHz 1850–1990 MHz 1710–1880 MHz	CDMA	QPSK down HPSK up	384 kbps 2 Mbps indoor

W-CDMA (TDD) (Wideband CDMA Time Division Duplex)	Korea, Japan, Europe, USA and Asian Countries	2003	1900–2025 MHz 1910–1930 MHz 1850–1990 MHz	TDMA/CDMA	QPSK down HPSK up	384 kbps 2 Mbps indoor
W-CDMA (low chip rate TDD) (Wideband CDMA Frequency Division Duplex)	Korea, Japan, Europe, USA and Asian Countries	2003	1900–2025 MHz 1910–1930 MHz 1850–1990 MHz	TDMA/CDMA	QPSK down HPSK up	384 kbps
TD – SCDMA (Time Division Synchronous CDMA)	China and other countries deploying W-CDMA (TDD)	2003	2010–2025 MHz GSM 900 band dcS 1800 band	TDMA/CDMA	QPSK 8 PSK	384 kbps phase 1 2 Mbps phasc 2
Bluetooth (Wireless Personal Area Network)	Worldwide	2000	2400–2483 MHz	TDMA	GFSK	1 Mbps now 2 Mbps future
802.11b (Wireless Local Area Network)	Mainly North America, eventually worldwide	1999	2400–2483 MHz	CSMA-CA	DBPSK / DQPSK	11 Mbps
802.11a (Wireless Local Area Network)	Mainly North America, eventually worldwide	2002	5150–5825 MHz	CSMA-CA	OFDM using BPSK / QPSK / 16 QAM / 64 QAM	54 Mbps
Hiperlan 2 (Wireless Local Area Network)	Mainly Europe, eventually worldwide	2002	5150–5825 MHz	CSMA-CA	OFDM using BPSK / QPSK / 16 QAM / 64 QAM	54 Mbps

2

Line of sight microwave systems

2.1   Introduction

As the term *line of sight* implies, this type of radio system developed from the idea that if you can see the distant end, then you can talk to it. This section of the book deals with point-to-point radio systems, which are usually setup on a permanent basis, to provide communication links between two or more fixed locations.

As discussed in Chapter 1, radio propagation waves are electromagnetic waves and as the frequency of the transmission increases from very low frequency (VLF) through the range up to extremely high frequency (EHF), the characteristics of the propagation through the atmosphere change significantly. For example, in Western Australia, there is a high power transmitter operating on a frequency of about 17 KHz – just above the audible range – and it can transmit low speed data to submerged submarines in most parts of the Indian Ocean. EHF systems on the other hand operate in the Gigahertz range, where frequencies are measured in thousands of Megahertz. These systems often have a range of only a few hundred meters and cannot *see* around corners or over hills.

As a general rule, the higher the frequency, the greater is the amount of information that can be transmitted over the system, so whilst an HF system on a ship in Auckland New Zealand harbor can reliably communicate directly with the ship's home office in Stockholm, it is limited to a single low-grade voice channel or very slow-speed data. On the other hand, a conventional digital radio system, operating in the 8 GHz band, can transmit digital data of 34 Mbps over 30 km and this can consist of voice, data and video information.

The range of frequencies used for point-to-point systems generally extend from 1 GHz up to the current limits of the band, at about 50 GHz. Above this, the spectrum moves towards the visible range where some links operate in the infrared region.

In this chapter the theory of radio link transmission and the components, which make up a system, are described. Some of the problems of radio transmission are discussed and an example of a simple radio path calculation for a radio link operating in the 1.7 GHz band, is given.

2.2   Background

Early point-to-point systems grew up with vacuum tube technology and by today's standards, were crude and cumbersome. Most systems operated in the VHF bands centered on 80 and 160 MHz. Aerial systems were large and limited in gain and to compensate for this, transmitters were often capable of delivering 100 watts or more. A typical system installed in the 1960s consisted of 250 V and 400 V dc power supplies, using selenium disc rectifiers, which occupied the lower third of a standard equipment rack. Next came separate transmitter and receiver units, which occupied about 30 cm of space, each. The transmitter had an RF output of 10 watts and often this was used to drive a power amplifier, to raise the output to 100 watts. There was a meter unit to measure the multitude of anode, grid, and cathode voltages and sometimes the luxury of an order wire circuit. Above all this was the frequency division multiplex equipment, which generally made use of early transistor technology, and at the top was an ac distribution system. This entire terminal rack had a capacity of four voice channels.

A modern radio terminal will be about the size of a facsimile machine and may have a capacity to carry 150 simultaneous telephone calls.

The rapid development of semiconductor technology allowed for an equally rapid growth in the use of point-to-point radio systems. Capacity increased up to 960 voice channels before digital systems began to force analog radios into the museum, while equipment size and cost per channel, decreased dramatically.

The past thirty years has seen a vast increase in the use of all types of communications, to aid the growth of developing countries, and to replace inadequate older systems in the Western world. It is thus relevant to look at the rise and decline of different types of communications links.

2.2.1   Overhead lines

Overhead open wire lines have all but disappeared. They are costly to install and maintain, vulnerable to damage and visually objectionable.

2.2.2   Underground copper cables

Underground broadband cable and coaxial cable systems were widely installed whilst labor costs were low and where the terrain was suitable. There systems could carry hundreds of analog channels but needed repeaters about every 10 km. Underground cables are vulnerable to attack from termites and burrowing animals. They can be damaged by careless plowing in agricultural areas and by excavations, in cities and towns. Like overhead lines, underground cables and their repeaters are vulnerable to damage by lightning, and repairs can be expensive and time consuming. Some underground cables were installed in pipes or ducts and this was to have an important consequence.

2.2.3   Radio

Line of sight radio offered many advantages over cable systems. In mountainous country, radio could be installed quickly and often cheaply from hilltop-to-hilltop whilst it was difficult to install cable in rocky hills and across rivers and lakes. Maintenance was required only at the radio site so the need to maintain access roads was reduced and while lightning was still a problem, it was possible to adequately protect radio sites. One important aspect of radio was, and still is, security. If necessary, radio sites can be protected with fencing and even with armed forces. Buried cables and repeater sites can easily be located and damaged by those hostile to the authorities.

2.2.4 Fiber optic cable

Fiber optic cable, has in recent years, offered a very significant challenge to radio. The large amounts of information, which we now transfer, require so much capacity that even modern digital radio systems are becoming insufficient. Fiber optic cable, today offers enormous capacity and an almost unlimited scope for further development. It is still vulnerable to accidental damage, sabotage and possible termite or animal damage but is not readily damaged by lightning. It can still be expensive to install but the huge capacity of multiple fiber cables can offset this expense and where old cables were installed in ducts, these ducts can be reused for new fiber cables. In some interesting cases, fiber optic cables have been pulled into disused oil and gas pipelines.

Today, fiber optic cable and radio systems both share the limelight and a careful analysis of the requirements, the terrain, and other relevant factors will lead the system designer to the correct choice.

	Digital capacity	Cost of eq for a 30 km route (US$)	Possible advantages	Possible disadvantages	Risk to system reliability
Radio	2 MB to 280 MB	100 K	• High capacity • Simple to install • Relatively easy to secure sites	• Interruptions due to extreme rain	• Lightning • Sabotage • Access to repeater sites
Fiber cable	2 MB to 280 MB (Note 1)	240 k cable 20K Eq. (Note 2)	• Very high capacity • Secure against storm damage	• High installation cost • May require land permits, etc	• Termites • Roadwork • Farming • Sabotage • Floods

Notes
1. *The digital capacity of a fiber cable is almost limitless because many fibers can be installed in the one cable for very little additional overall cost.*
2. *Cable costs vary according to construction and protective coatings.*

Table 2.1
Some comparisons between radio and fiber cable systems

2.3 Point-to-point radio systems

For many years, point-to-point links have been the main application of fixed radio systems. They provide communications circuits between one fixed point and another. In public communications, towns and cities are linked together, in industrial situations a central plant may be linked to RTUs at remote mines or bore sites and in military situations transportable terminals are used to link a central command location, to a forward position.

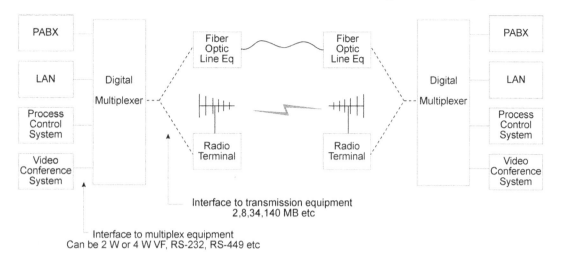

Figure 2.1
Typical digital radio or fiber cable system

If nature is kind to the system planner, a single hop can often link one town to another or perhaps to an industrial complex, which lies up to 40 km away. If the terrain is mountainous it may be necessary to use the configuration shown in Figure 2.2, which has one hop to go from the first town up to a hilltop repeater station, another to traverse a valley and a third to go from another mountain to the second town below. Where large towns or cities, perhaps hundreds of kilometers apart, are to be linked and a high volume of traffic justifies the cost, a system may use many repeater stations, each separated by 50 or so kilometers. Sometimes smaller so-called 'spur' links from the repeater stations may service towns along the route.

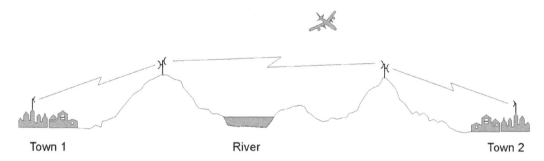

Figure 2.2
A short radio link traversing a mountain range

A typical application of point-to-point radio is shown in Figure 2.3 and it is used to provide a variety of communications services for railroad and pipeline systems. A series of repeater stations will be located along the route, usually spaced 30 to 50 km apart. The radio link will often carry telephone and data circuits, to mines or petroleum fields located at the distant end. At each repeater site, some circuits will be dropped out of the link to connect into mobile radio base stations, which will provide good communications for maintenance staff working on the railroad or pipeline. Other drops will be used to relay railroad signaling information or pipeline telemetry data to wayside stations. Some railroad operators use wayside hotbox detectors, which contain an infrared sensor, mounted close to the passing wheels of rolling stock. When an overheated bearing is detected, the device can signal the details to the railroad controller via the radio system and possibly avoid a serious accident, caused by a seized bearing. Still more drops will provide telephone and data connections to wayside maintenance camps.

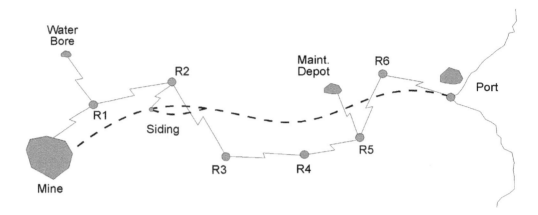

Figure 2.3
A long radio link serving a mining operation

2.4   Point-to-multipoint

There are some situations, which pose special problems for the system designer. Consider the case of a town surrounded by ten or twenty very small communities, each with very limited telephone or data needs, or a central control station, which monitors the operation of many remote bore pumps, reservoirs or electrical switching stations. The designer could specify multiple point-to-point links, one to each site, but this could result in a very large and expensive installation at the central station (requiring an enormous number of antennas on a single mast). Most of these links would be used for perhaps only a few minutes each day so the whole system would be extremely inefficient. Another problem would be the use of ten or twenty radio frequency pairs, as each link would require a separate frequency. Consider also the problems, which were discussed in Section 1.15 of intermodulation and interference.

A more practical solution lies in systems, which use a type of time division multiplexing known as time division multiple access (TDMA) or sometimes as demand assignment multiple access (DAMA), similar to those discussed in the chapter on satellites.

TDMA systems were developed in the 1980s and the early developments were in analog radio. In this type of system, a single central station transmitter interrogates each remote station sequentially by means of a code. At this stage, all the remote receivers are on and all the transmitters off. If a particular station requires a connection to the central station, the transmitter at that station will respond with a reply code and the central station will stop scanning and setup a full duplex connection to the remote site. A data or telephone call may then proceed and when completed, the remote transmitter will turn off and the central station will begin scanning again. When a call originated at the central station for a remote site, the code transmitted to the required remote contained an additional instruction to turn the remote transmitter on so that a call could be setup. These systems were simple and effective but were limited to a single call and if a call were in progress, other remote sites had to wait until the line was free again.

TDMA systems using digital technology are now preferred because of their better performance and their ability to integrate directly into digital transmission systems.

A typical TDMA system operates in the 1.5 or 2.5 GHz band and can provide 30 simultaneous traffic circuits each running at 64 kbps. It can handle up to 158 remote stations and 480 subscribers, each of which will have a dedicated telephone number, so that the system can be fully integrated into any normal telephone network.

If a location is too far removed from the central station, for a direct radio link, a repeater station can be used to extend the range. If required, the repeater can have subscribers attached to it or if there is no traffic requirement, it can operate as a straight repeater.

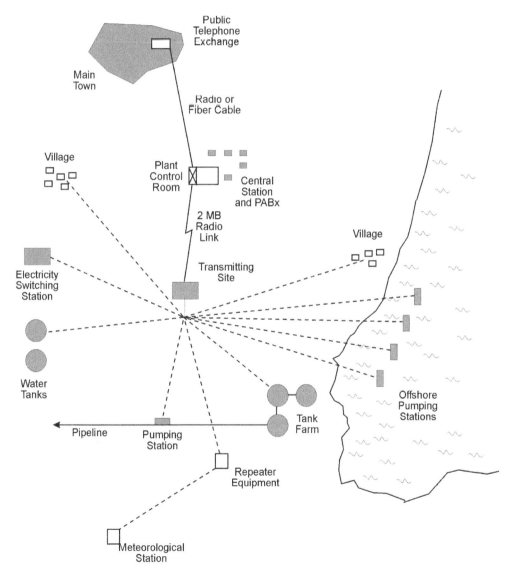

Figure 2.4
A point-to-multipoint system

Figure 2.4 shows a block diagram of a typical TDMA system, which could provide voice, and telemetry services to offshore wellheads and similar services to onshore tanks, pumps, and water and electricity installations. It also serves a nearby village where telephone services could be provided to plant workers as well as the public facilities such as a medical center, police, post office and the like. It is possible to setup telex services and even payphones to operate over the system. It is also possible to setup permanent low speed data circuits, which operate between 1200 bps and 19.2 kbps, between the plant control room and remote sites or between two remote sites.

Note that the TDMA central station is located in the plant control room with the plant PABX whilst the transmitting equipment is installed at a hilltop site where it will provide the best coverage of the remote sites. The connection between the two types of equipment is a normal 2 MB link, which would probably be a radio link because of the terrain. The central station would be connected to the PABX by as many telephone pairs, as there are subscribers, on the TDMA system

If there were a nearby local town with a public exchange, it would be connected to the PABX, by radio or cable in the normal way, and by switching through the PABX as many of the remote sites as were required, could have access to the National public network. Data services to the remote sites would be switched by the PABX or connected directly to a router, as required for onward connection into the plant data system.

2.4.1   Theory of operation

The theory of TDMA applicable to satellite systems is discussed in section 3.4.2.2.

The way in which a typical microwave TDMA system operates requires an understanding of the digital multiplexing process, which is described in Section 2.8, and it is recommended that the reader understand this section first before proceeding.

Figure 2.8 shows how PCM coded voice information and/or basic rate 64 kB data can be stacked in a 2048 kbps timeframe. There are 32 time slots available and two of these are used for signaling and supervising the 30 remaining time slots that can be used for the system traffic. The central station transmits continuously and all the remote sites receive all the information, but each site decodes only that time slot which carries its unique address. In this way it is possible for the central station to setup the transmission of a   64 kbps stream to each 30 separate location or alternatively, 10 streams to 1 location, six to another, two streams each to another four locations and the remaining six to separate locations.

Each remote station knows which time slot it decoded when it received the incoming transmission and in order to transmit back to the central station, the remote station turns on its transmitter at precisely the beginning of that particular timeframe and turns it off at precisely the end of the timeframe. In this way, as before, all 30 remote sites can transmit on the same frequency to the central station.

Because the distance from each remote site to the central station will not be the same, the transit time of the signals to and from the remote will differ. The central station is able to monitor these variations and constantly adjusts for minute timing errors. Some systems are now so accurate in this timing control that they are able to maintain communication with a slowly moving vehicle, such as an excavator, or reclaimer, and a vessel such as a drill ship or supply boat.

The above is a brief explanation of the operation of a typical TDMA system and in practice the timeslot arrangement and signaling, are rather more complex than has been described.

2.5 A typical radio terminal

The operation of a radio transmitter or receiver need not be of any deep concern to the system designer who will normally be able to regard them merely as black boxes and concentrate on the overall specifications.

Figure 2.5 illustrates a simplified block diagram of a digital radio terminal and this section will briefly discuss the function of the component modules.

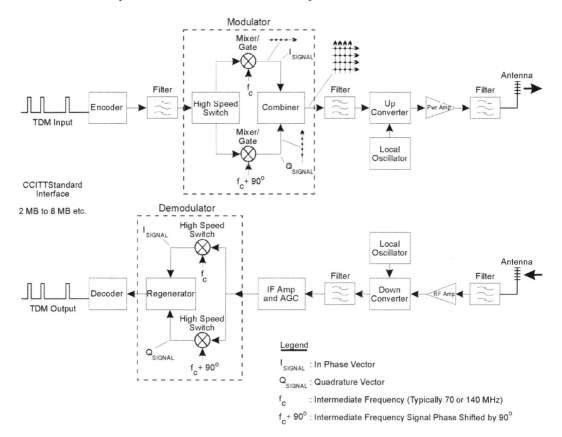

Figure 2.5
Block diagram of a digital radio terminal

2.5.1 The transmitter

A digital signal coded to CCITT standards is fed to the encoder, where it is buffered and has bit stuffing and parity information added. The signal passes through a band-pass filter to remove noise and harmonic products. It is then passed to the modulator where a high-speed switch directs, in this case binary information, to one of two mixer or gate circuits where one or the other gate will be enabled according to the binary state. One gate feeds a carrier signal at an intermediate frequency of 70 or 140 MHz. The other gate feeds the same frequency, with a phase shift by 90°, to another band pass filter. This feed may be to another mixer or up-converter, where the intermediate frequency is mixed with a local oscillator. This local oscillator will run at a frequency, 70 or 140 MHz, below the required radio frequency.

The resultant radio carrier will then be fed to a power amplifier to increase the power to perhaps 1.0 to 10.0 watts and after passing through yet another band pass filter, it will pass through the antenna feeder cable to the antenna.

2.5.2 The receiver

A radio frequency signal, received by the antenna will enter the receiver and pass through a band pass filter, to remove other interference, which may have been collected by the antenna. It then passes the signal to an RF amplifier and to a mixer or down converter where, as in the case of the transmitter, it will be mixed with a signal from a local oscillator, which differs from the radio frequency, by the intermediate frequency of 70 or 140 MHz. The output from the down converter will be at the intermediate frequency and is fed to an amplifier. This IF amplifier is able to compensate, to a large degree, for variations in the RF input signal level at the antenna, caused by signal fading, so that the input may drop by 50 or 60 dB whilst the output will remain almost constant. This automatic gain control (AGC) is a very important element in maintaining the stability of the system, over a wide range of operating conditions.

The IF signal then passes to the demodulator, where it divides to feed two high-speed switches or phase detectors. When the IF signal is in phase-one switch (in the diagram), it will generate a binary 0, whilst the other will generate a binary 1, when the IF signal is in phase-quadrature. These states are fed to a regenerator, which reconstitutes the digital signal, and after passing through a decoder, which performs parity checks and converts the signal to a CCITT format; the signal is fed from the receiver output to the multiplex system.

The above will serve to provide only a broad outline of the principles of a digital radio system. In practical systems, modules that are more peripheral will be found and more complex modulation methods such as multilevel quadrature amplitude modulation (M–QAM) will be used to allow the modulation process to be more efficient and so to allow high-speed data streams to be passed over the radio link.

2.6 Modulation methods

The digital modulation and demodulation system, shown in the previous section, is of the very simplest type and in itself is very inefficient and capable of only slow-speed data rates. The signal fed to the modulator is a binary signal, consisting of two voltage levels or signal states and from this basic information, a digital word is built up to represent the sampled voltage of the input signal. In the previous section it was shown that the intermediate frequency and hence the carrier frequency was shifted in phase by 90°, to represent the change in state of the binary signal. This can be represented as a vector and is illustrated in Figure 2.6 as 1-PSK.

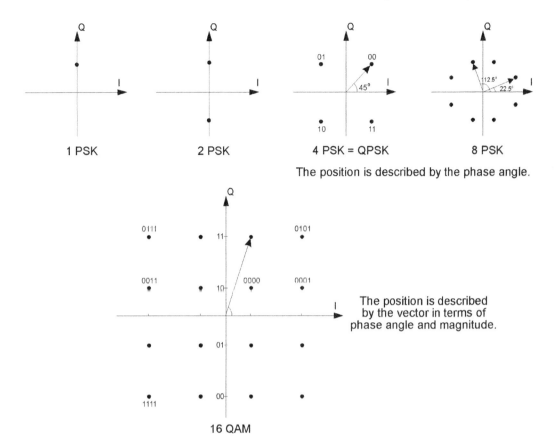

Figure 2.6
Some modulation patterns

Following on from 1-PSK, it would be easy to shift the phase both + and –90° and these two conditions could be represented as 2 PSK. Each of these conditions could denote a specific voltage level on the sampled input signal.

The next step is to shift the phase in smaller increments, say 45° and now the carrier signal can be phase-modulated to denote four conditions, by moving the phase to 45°, 135°, 225°, and 315°. Each of these four states can now be used to directly represent a two-bit word as shown in the four PSK or QPSK diagram.

By shifting the phase, only 22.5° the pattern can be increased to 8 PSK and the throughput of the modulator is three times that for a 2 PSK system. If the pattern is increased to 16 dots, the simple phase shift angle will not be sufficient to differentiate the pattern in the demodulator. However, the same pattern can be generated by a type of amplitude modulation called quadrature amplitude modulation and because QAM defines the dot position as a vector in terms of phase-angle and amplitude, it becomes a simple matter to generate 16 QAM, which in turn allows for a four-bit word.

From the last step, it is now possible to go on to 64 QAM and even higher levels, which greatly increase the throughput of the modulation and demodulation stages. The circuits, which perform these functions, tend to be quite complex and both difficult and unnecessary for the system designer to understand because he will generally be using standard transmitter and receiver modules, to handle the data rate required.

2.7 Standards

There are many organizations, which set standards that are relevant to the electronics and communications industries. Some are national in their extent whilst others such as CCITT (Consultative Committee for International Telephone and Telegraph) develop standards for use on a worldwide basis. (The CCITT has now been incorporated into the International Telecommunications Union (ITU).)

Most national public carriers set their own standards for interfaces to private subscriber equipment and whilst these are generally based on common international standards, they are often modified to suit local conditions.

Whilst international standards may seem to be of little interest to local system designers, they become very relevant if their network is connected into the national network and hence the international networks. Consider the case where an engineer at a remote plant site in Australia calls over a private network into the national network and thence to a factory in Poland. There are many tandem sections in such a call and unless each section maintains strict limits to the permitted degrading of the call, it will be so distorted as to be useless.

In analog systems, the problems of maintaining adequate speech levels, against the cumulative distortion fed in at each stage of modulation and demodulation, placed severe limitations on the system designer and substandard systems were not permitted to connect to the national network. The situation with digital systems is much easier because a digital signal can be regenerated many times, without degrading the information, where bit-error rates define the performance of the system. Unfortunately, some distortion is introduced in the analog to digital and subsequent digital to analog conversion of a voice circuit and this quantizing distortion will be discussed in section 2.8.2. The important point to note is that each A–D and D–A conversion introduces a defined number of quantizing distortion units (QDUs) and once the private network exceeds the permitted number of QDUs, interconnection to the national network is not permitted.

An important set of standards is used to regulate the interconnection of national networks and of many other sections of the communications world. These standards are issued by the International Telecommunications Union (ITU). The International Consultative Committee for Radio (CCIR) series of regulations, cover many aspects of radio communication and the CCITT series covers many aspects of telephone and data communication. Whilst the title is a little dated, this series covers most modern aspects such as data, facsimile, and video communications.

Each country has a regulatory body that publishes interconnect standards for connection into public networks. Generally, included in these is customer premise wiring standards and licensing requirements.

It is very important that the system designer be aware of these regulations and standards, which are mandatory for many interconnected networks, and provide very useful criteria for the specification of almost all types of communication equipment.

2.8 Multiplex equipment and data rates

2.8.1 Analogue multiplex systems

Analogue multiplex equipment is now rarely used in modern equipment. It used a frequency division multiplex (FDM) system. Twelve analog voice channels were combined in a frequency separated stack known as a group and five of these groups again combined to form a supergroup of 60 channels.

Depending on the capacity of the system, these supergroups were again combined up to the maximum of 960 channels. Whilst the system was reliable, the need to transmit digital information, without the use of modems, rapidly made analog transmission obsolete.

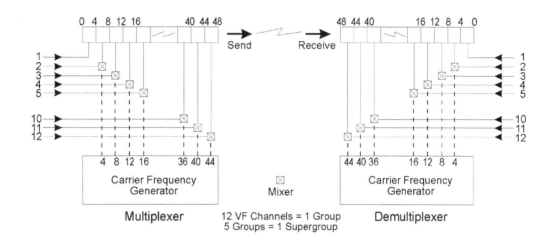

Figure 2.7
A basic frequency division multiplex system

Figure 2.7 shows a block diagram of a simple FDM system. Twelve incoming voice channels are fed to a group of mixers, where eleven of them combine with the multiple outputs of a carrier frequency generator. The VF bandwidth of the channels is limited, usually within the range of 300 to 3 400 Hz. Between 3.40 kHz and 3.8 kHz, there are two single-bit data channels, which are used for the E and M signaling, used in older type telephone signaling systems. The space below 300 Hz and above 3.8 kHz is kept free as guard bands, to prevent interference with adjacent channels. The first channel is not mixed and enters the base-band as the lower frequency slot. The second channel is mixed with a 4 kHz carrier and the sum product is a frequency band from 4.3 to 7.4 kHz, with the E and M signals in the band from 7.4 to 7.8 kHz. The third channel is mixed with an 8 kHz carrier and is multiplexed into the third base-band slot between 8 and 12 kHz and so on up until the twelfth channel, which is placed into the 44 to 48 kHz slot.

At the receiving end, the process is reversed but instead of using the sum product of the mixing process, the difference product is used and the twelve multiplexed channels in the base-band are de-multiplexed into twelve separate VF channels.

Usually, the same carrier generator is used for MUX and DEMUX and it is important that the carrier generators at each end of the system are synchronized.

2.8.2   Digital multiplex

Before looking at digital multiplex systems it is useful to have some understanding of the way analog signals are converted into digital format and Figure 2.8 illustrates how this is done using pulse code modulation (or PCM).

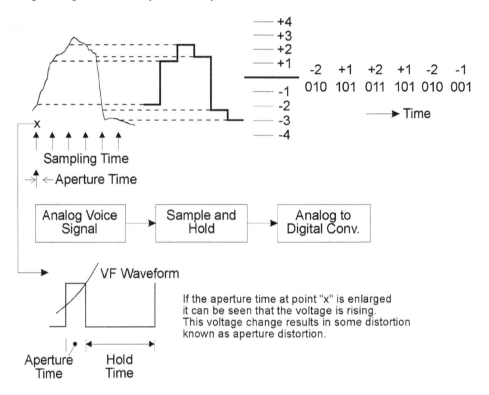

Figure 2.8
Basic pulse code modulation

The amplitude of the analog waveform is sampled at a regular rate and the sampled amplitude is maintained by a hold circuit until the next sample interval. These voltage levels are rounded, up or down, to the nearest of 128 or 256 discrete voltage steps and these steps are given a digital number of 7 or 8 bits, representing their amplitude.

If a signal is sampled at a rate that is twice the maximum frequency of the signal, the original can be reproduced in an acceptable form from the samples (Nyquist sampling rate). If a 0 to 4 kHz VF signal is to be transmitted, the sampling rate will be twice 4 k or 8000 samples per second and if each sample is an 8 bit digital number the bit rate will be 64 kbps per second (8×8 k).

This 64 kbps unit forms the basis of most current digital modulation schemes that use the European CCITT standards, whilst the American system sometimes uses a 7 bit sample, which results in a 56 kbps unit.

Whilst a digital signal is not distorted by noise, and can easily be regenerated as often as required in a transmission system, there is some distortion introduced by the A–D conversion, and the subsequent D–A conversion at the receiving end. This is called quantizing distortion and results from the limited number of samples which can be taken of the analog waveform, so that instead of reproducing a pure waveform, the result is a little jagged, something like a hacksaw blade. This limits the number of times a signal may be converted from A–D and back again. These conversions are assigned a value of quantization distortion units (QDU), which in turn are used to project the end-to-end performance of a digital transmission system.

Beginning with the base 64 kbps unit, a group of 30 channels can be placed 'end-to-end' to form a multiplexed bit stream, which runs at 2.048 Mbps. This is known as Level 1 multiplex or 1 DME and actually contains 32×64 kbps. According to the CCITT design, the first time slot is used to carry synchronization and alarm information and time slot 16 is used to carry all the E and M signaling information for the 30 VF channels. Figure 2.10 shows a model of this structure. Internationally this is often referred to as '**E1**'PCM technique.

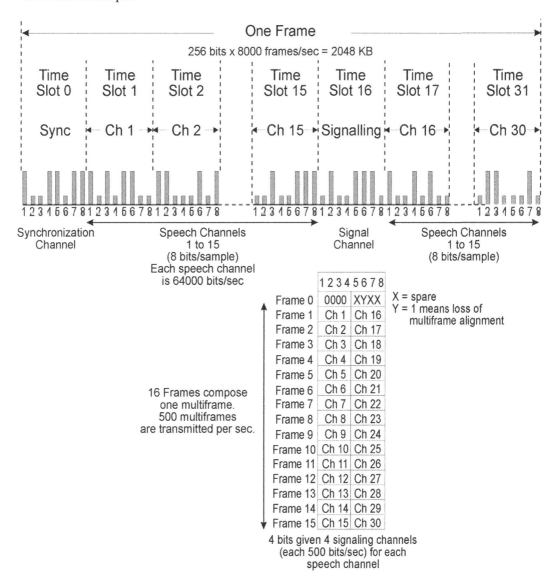

Figure 2.9
CCITT model of 30 × 64 kB speech channels in a 2.048 MB stream

The basic 2.048 Mb stream, which is commonly referred to simply as 2 Mb, can then be merged or multiplexed with three other identically framed 2 Mb streams, to the 2 DME level of 8 Mb and so on, to level 3 DME, which is at 34 Mb, rather than 4 times 8 or 32 Mb. The extra 2 Mb is used to allow the lower order bit streams to float around a little, if their synchronizing or timing information is not exactly the same. The 34 Mb streams can be multiplexed to even higher capacity systems that can run at very high bit rates. Such systems are common on national and international circuits and run at data rates of 140 and 565 Mbps, but for most industrial application systems, running at 2 or 8 Mbps, is the norm.

North America developed a multiplexing technique in the 1960s, before the CCITT **'E1'** international standard was developed. Because of the early introduction and its widespread use and acceptance, it has been promulgated as the American standard for multiplexing. It is often referred to as **'T1' PCM**.

The 'T1' technique is similar to 'E1' in that speech is sampled with 8 bit resolution, but this time there are only 24 time slots per frame. Therefore, conforming to the Nyquist theorem and sampling 8000 times per second, each voice/data channel will have a data rate of 64 kbps. All 24 slots are used for voice or data. Therefore the frame speed is 24 × 64 bps = 1.544 Mbps (compared to 2.048 Mbps for 'E1').

Earlier techniques required that the eighth bit of every time slot is used for signaling information. This is shown in Figure 2.10 below.

This is obviously very inefficient as one eighth of the information sent is used for signaling. It also effectively reduces the sampling resolution down to 7 bits per time slot, and the data channel speed to 56 kbps. This can degrade a voice signal down to levels below acceptable telephone quality.

Modern 'T1' equipment overcomes some of these limitations by sharing the signaling bits between channels and by only sending signaling in certain frames (i.e. when it is required).

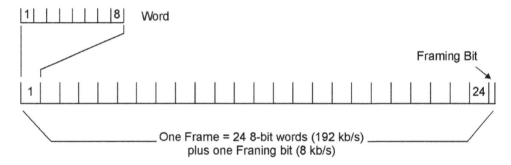

Figure 2.10
A TI frame

The CCITT EI standard is slowly becoming the accepted standard throughout the world and is finding its way into the North American market. It is also worth noting that 'E1' PCM is the standard on which ISDN has been developed.

The scheme described above was the first development in digital multiplex technology. At first, the system was simply known as digital multiplex. Since the capacity of fiber optic cables allowed high-speed traffic, new types of multiplex systems have been developed. The original system was called by the almost unpronounceable name of **Plesiochronous transmission**, to distinguish it from other, more recent types of systems, which are described later in this section.

2.8.3   Voice compression

The PCM method of A–D conversion is capable of reasonably good quality transmission of a wide range of VF signals, including music, albeit with a limited frequency range. However, it is fairly inefficient in terms of the data rate used to carry speech, which is after all the main component of telephone traffic. Therefore, PCM is really quite an unsophisticated coding system, considering the amount of data that can be carried by a 64 kbps stream. An alternative system known as Adaptive Differential PCM (or ADPCM) has become widely used. Instead of measuring the actual amplitude of the sampled waveform, ADPCM takes advantage of the fact that consecutive samples will be quite close in value and so the change in amplitude is encoded. By using ADPCM, a significantly improved performance, at 64 kbps, is possible or alternatively, at 32 kbps the performance is equal to 64 kbps PCM.

Many more ingenious methods of encoding are being developed and many manufacturers now offer encoding at 8 kbps and even lower. However, the quality is usually not acceptable for public communications but may be quite adequate for service channels or other engineering circuits.

2.8.4   Synchronous digital hierarchy

In the preceding paragraphs it was noted that the timing of 2 Mb streams varies a little within prescribed limits. Each lower order multiplex stream has its own clocking source, which causes problems when fed into a higher speed stream. In fact this system is asynchronous and in the higher-level multiplex stages they undergo bit stuffing, which adds as many bits as required to bring each individual stream up to a standard speed and then these bits must be removed at the demultiplexing stage. Whilst this process need not concern the designer working with small 2 Mbps systems it does become a big problem for national and international trunk network designers, particularly when 2 Mbps streams are to be dropped and reinserted along a radio or fiber cable route.

In the late 1980s, a system was developed which uses synchronous transmission and because every 2 Mbps stream is precisely synchronized, they can be directly inserted into a high speed bit stream and dropped from it as required with a minimum of equipment. Because some applications require very high data rates, there are provisions to drop and insert or '*map*' bit streams up to 10 Gbps.

The synchronous digital hierarchy (or SDH) used in Europe and Asia and synchronous optical networks (SONET) used in America are also designed to accommodate both the European and American PDH structures. Therefore, the recognized lower order tributaries to the main higher data rate streams are based on 2-8-34 and 140 Mbps (SDH) and 1.5-6 and 45 Mbps (SONET).

The common SDH/SONET data rates are:

- STM 1/OC 3   155 Mbps
- STM 4/OC12   622 Mbps
- STM 16/OC 48   2488 Mbps
- STM 64/OC 192   10 000 Mbps

Microwave links are readily available and well priced that operate at STM 1 speeds. Single link STM 4 microwave links are also available, but these are significantly more expensive and complex.

2.8.5   The 2 Mbps world

For the system designer who needs digital transmission there is a wide range of input-output facilities available. In a typical system, there will be a 2 Mbps line interface, which can connect directly into a radio link, a fiber transmission system or a higher order multiplex. The interface can serve as a terminal unit or to provide a drop and insert access to a repeater station. It provides all the synchronization and supervisory information to allow communication with distant line units. Figure 2.11 illustrates the general arrangement of a typical system.

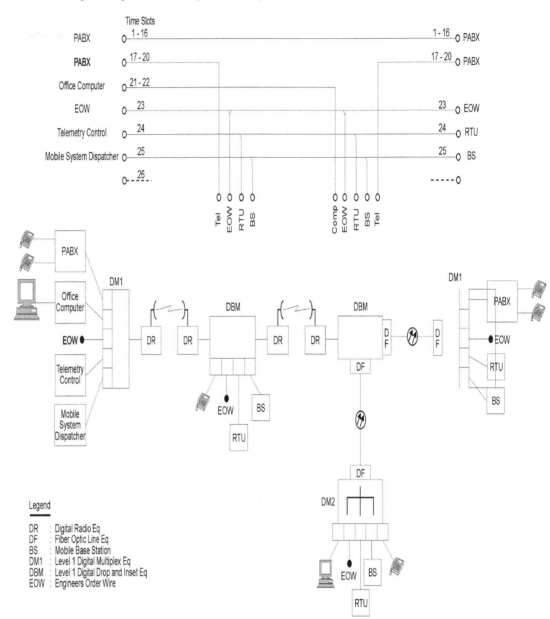

Figure 2.11
Example of a 2 Mbps system using radio and fiber cable transmission circuits

The input–output units are installed into the line interface as required. Depending on the complexity of the circuit, there will be up to ten circuits per I-O unit and this gives the designer scope to connect almost any type of equipment. Figure 2.12 gives an idea of the type of interfaces that can be installed into a line interface.

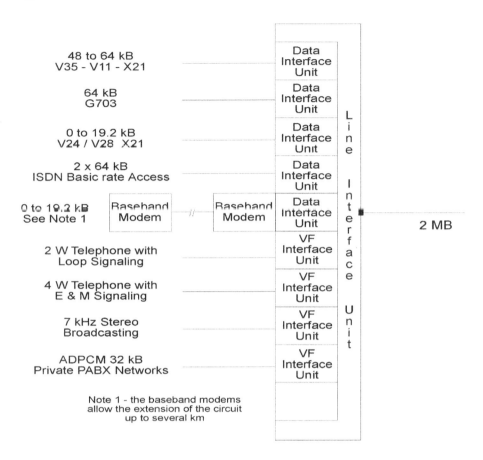

Figure 2.12
Some of the interface units, which may be used with a 2 Mbps, line interface

2.9   Antennas and multicouplers

Parabolic or dish antennas are generally used for frequencies above 900 MHz, although, Yagi types are still common at that frequency. Above 900 MHz, the capacitance and inductance of the elements in a Yagi, become limiting factors and manufacturing becomes difficult and expensive.

The parabolic antenna consists of an illuminator, which is generally a dipole and one that is set at the focal point of the reflector dish. The antenna operates exactly as a searchlight would, by accurately focusing most of the energy from the illuminator into a narrow beam, or, in the case of a receiving antenna, by collecting the energy falling on the reflector surface and focusing it onto the dipole, where it is converted back into electrical energy. The size of the antenna is determined by the operating frequency, where the lower the frequency, the larger the diameter of the dish, and by the gain required, where the higher the gain, the larger the dish.

In radio-link systems, antenna beam-widths are in fact quite narrow due to the very high gain of parabolic antennas. The table below illustrates this and it is evident that these antennas must be pointed to the distant end with great accuracy.

ANTENNA BEAM-WIDTHS	
Gain in dB	**Half power beam-width**
30	5°
35	3°
44	1°

Table 2.2
Antenna beam-widths

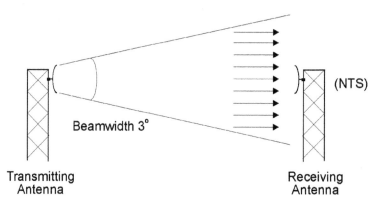

Figure 2.13a
Antenna beam-widths

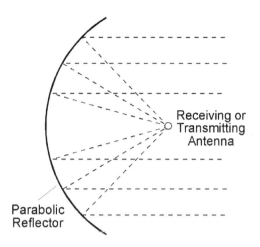

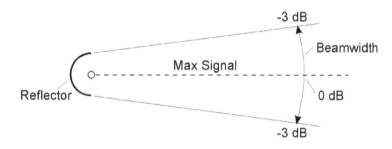

Figure 2.13b
Illustration of a parabolic antenna and its beam-width

The beam-width of the dish is determined by measuring the power levels either side of the azimuth, until they have dropped to half power (i.e. by 3 dB). From one 3 dB point to the other, is then, the beam-width.

2.9.1   The use of reflectors

Sometimes a hill, even a large building, or a tank, will make it impossible to obtain a line of sight path even though the path may be quite short. A very typical case is a ski village built at the bottom of a ski run off a mountain. To avoid the cost and complexity of a normal repeater station at the top of the mountain, the system designer can sometimes use a large metal reflector, which is used in exactly the same way as a mirror would be, to reflect the radio beam aimed at it from the village below, to the distant receiver site and vice versa. As with a mirror, it would not be possible to cover certain reflection angles, such as those near to 180°, and in such cases, two reflectors can be used to give a double reflection of the beam. Such an arrangement is shown in Figure 2.14.

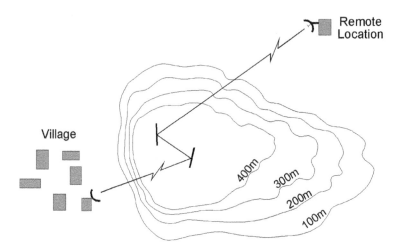

Figure 2.14
A passive double reflector

At frequencies above 6 GHz, reflectors can become extremely efficient if the surface is very flat and that they are sufficiently large. A typical reflector would be about 6 m by 9 m and this would reflect almost 100% of the incident radio wave.

2.9.2   The use of passive repeaters

At frequencies below 2 GHz, passive reflectors tend to become very large and therefore impractical. At these lower frequencies, it is possible to install two parabolic antennas – one pointing to the nearby station and the other to the distant terminal. A short piece of coaxial cable connects the two antennas together and the signal received by the first antenna is retransmitted by the second. Unfortunately, this system is not as efficient as the reflector system, because it behaves as two paths in tandem rather than one single reflected-path. Tandem path losses must be carefully calculated and may be too high to allow the system to operate. Figure 2.15 shows how a passive repeater would be used.

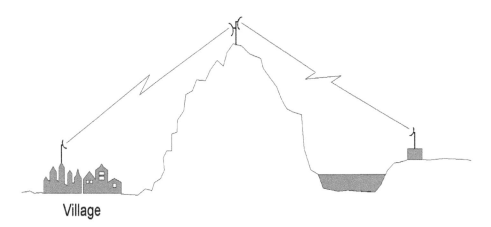

Figure 2.15
A passive repeater

Note that both passive repeaters and reflectors are generally, only satisfactory when they are close to one end of the path.

2.9.3 Duplexers and multicouplers

After reading the preceding section, one will appreciate that antennas can sometimes prove to be very expensive items. Not only can the antenna and feeder cable amount to thousands of dollars, but a supporting structure, with its associated wind loading design, can cost many times more than the antenna, especially if a tall mast or tower has to be built in a cyclone prone region. Consequently, every effort should be made to use it to the greatest advantage.

Sometimes several systems in the same frequency band need to serve the same general area. Perhaps a point-to-point link has no spare capacity and a second link needs to be installed. Perhaps a hot standby backup is added to an existing link, to improve reliability, or perhaps several independent mobile radio systems are required to cover the same area. In situations like this, it is possible to allow several transmitters and receivers to share the same antenna.

Section 1.15 covered the use of multicouplers and duplexers and the use of these devices in point-to-point systems is very similar.

2.10 Coaxial cables and waveguides

The theory of waveguide transmission is complex but the very name **waveguide** yields a simple explanation. At the transmitter, the electrical energy is fed via a short cable to an illuminator, which is again rather like a small dipole accurately located within the waveguide. This is nothing more than a precisely drawn rectangular copper tube.

The electrical energy is radiated into the waveguide and because the dimensions of the tube are a function of the wavelength, the radiated energy wave is reflected off the walls of the tube and guided along the length of the tube, to the antenna. At the antenna, the signal is sometimes converted back to electrical energy by an identical dipole illuminator, and connected to the antenna by a short cable, but more often than not, the waveguide runs to the focal point of the antenna and is directed towards the reflector through a launcher. This is rather like an exponentially tapered rectangular trumpet.

Rectangular waveguide is expensive to manufacture, delicate to handle – because any irregularity or dent will disturb the signal, expensive to install – because it needs precision bends to go around corners, and vulnerable to moisture ingress through the joints that connect it all together.

To overcome some of the problems of rectangular waveguide, some manufacturers developed a form of waveguide, which uses the same type of spiral copper construction that is used for larger coaxial cables, to make a very accurately dimensioned oval tube. The center point of the tube follows an ellipse or spiral so that at any point along the length of the guide, the diameter and the profile remain constant.

This elliptical waveguide is relatively easy to transport, handle, and install. It is covered with a tough protective sheath underlaid with a spongy tar-like compound, to absorb moderate rough handling and is completely weatherproof. Due to the absence of joints and bends, an elliptical waveguide often provides a lower loss transmission feeder than a rectangular waveguide.

2.10.1   Avoiding the use of waveguides?

It was noted in the introduction to Chapter 2 that the frequencies used for point-to-point links extend up to around 40 GHz. At these frequencies, the attenuation of coaxial cable is extremely high and even waveguide losses can become unmanageable.

There is a growing demand for point-to-point links that can be used to cover short distances and this is particularly so in city areas. Consider the case of a plant, which expands to new land across a main road from the existing site and where the costs of putting a cable across the road would be very high. At other times, a temporary link may be needed for a special requirement such as a construction program. The frequencies, in the upper part of the band (over 20 GHz), are very suitable for these applications because the small antennas used are quite directional and the high free space attenuation of the transmitted signal allows many links to be used in a city without a great risk of interference.

To overcome the problems of high feeder-loss, the transmitter/receiver unit is usually contained in a very robust and weatherproof metal case. The parabolic antenna is mounted directly onto the front of the case so that the feed horn and duplexer are usually a single-assembly, with very short connections to the receiver and transmitter. In cold climates, the case will be electrically heated and some form of double skin or sunshade used, to keep the equipment within a tolerable operating range.

A composite cable, which carries the data input and output signals as well as the operating and heating power, connects the compact unit from its location on a rooftop or antenna support structure, to the multiplexing equipment, usually located in an equipment room.

These mast head units, as they are generally known, find many applications where their relatively short range of 3 to 7 km can be used to great advantage. The main disadvantages lie in the difficulty to service them when required, as it may require a rigging team to bring them down from their mounts, their vulnerability to damage by cyclonic winds (in some areas), and to theft or malicious damage.

2.11   Power supplies

Most modern telecommunications equipment operates from dc supplies. This makes it a relatively simple matter to plan battery backup systems, to ensure reliability.

The common supply voltages for radio and multiplex equipment are –24 and –48 Volts. These are very practical voltages for a dc system, as they are low enough to be relatively safe in case of accidental contact with the supply line, and high enough to avoid the high currents involved in lower voltage installations. In many cases, modern equipment can operate over a wide range of voltages and so fewer problems should be encountered when charging batteries. Some equipment, which uses ac–dc converters to provide the internal operating supplies, will operate over a range from about 22 volts up to 72 volts.

Some equipment, which may be derived from a mobile radio background, operate from +12 volt supplies and this can be a disadvantage where they are installed with equipment operating from –24 or –48 volt supplies. Where both installations are substantial, it may be necessary to install two separate battery supplies. If the site has a majority of +12 volt equipment it may be practical to install 12–24 or 12–48 converters, to supply the minority equipment and similarly, if the 12 volt load is low it may be better to use down converters than to install a separate supply.

Battery systems and primary power supplies will be discussed fully in Chapter 5.

2.12   Path loss

2.12.1   Free space attenuation

Line of sight path loss or attenuation was briefly mentioned in Section 1.4. This section looks at the factors that cause path loss, and leads to Section 2.13 where a simple case study of a path will be made.

When a radio signal passes through the atmosphere, from a transmitting antenna to a receiving antenna, it will be attenuated – as a function of the frequency of the signal and the distance between the antennas. This is called the free space loss and it occurs because the power of the signal is reduced by 6 dB for every time that the distance from the antenna is doubled.

Once again, the formula for calculating the free space attenuation between two dipole antennas is:

$$A = 92.4 + 20 \, log \, D + 20 \, log \, F$$

Where:

    A is the free space loss in dB
    D is the distance in kilometers
    f is the frequency in GHz

Where the distance is measured in statute miles, the constant 96.6 should be used instead of that shown above.

2.12.2   Rain attenuation

At frequencies below 8 GHz, the attenuation due to rainfall is low enough to be considered insignificant. However, above 10 GHz the effect of heavy rainfall becomes very significant and must be taken into account when planning a system. Figure 2.16 shows the relation between rainfall and attenuation.

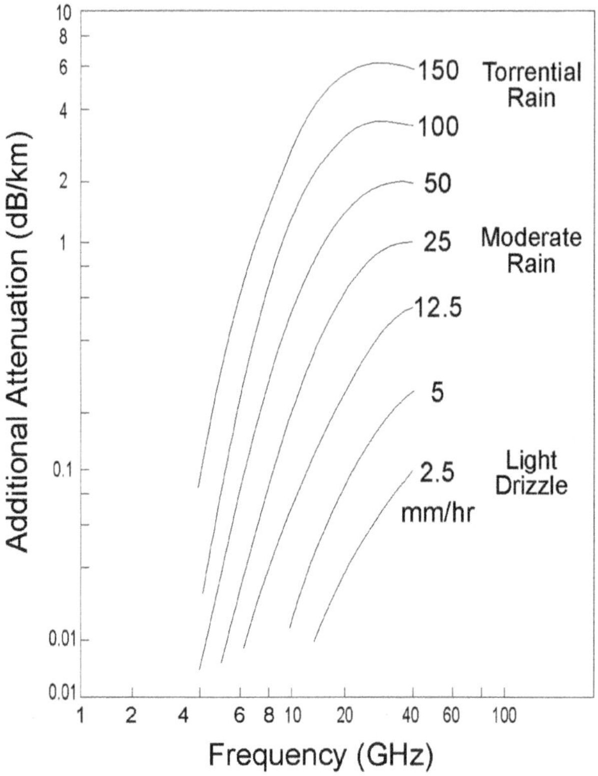

Figure 2.16
Attenuation due to rainfall

2.12.3 Fade margin

It could be supposed that, if the free space loss of a path were calculated and due allowance made for rain attenuation, the free space loss story would end. However, there are some other factors that need to be allowed for. The most prominent of these will be multipath propagation, which is discussed in the next section. Sometimes transmitter power or receiver sensitivity will degrade with age or due to faults. Sometimes antennas will be damaged by moisture ingress or strong winds. Sometimes birds will nest in the antennas and their droppings may cause corrosion.

To allow for all these unplanned losses, it is prudent to ensure good allowance for degradation, in the path calculation. A common figure to allow is 30 dB. On very long hops or paths over water where very severe fading can occur, it is advisable to allow a higher fade margin.

2.13 A simple path calculation

This section will bring together some of the topics covered in previous sections and these will be used to develop a path loss calculation. This relates to an imaginary path between station *A* and station *B*, on a railroad communications system 43 km apart.

The two sites are fixed, as there are existing buildings with masts at both sites, and the aim is to verify that the path once used for a UHF link will be suitable for a new 2 Mb digital link.

PATH LENGTH – 43 KM	STATION A	STATION B
Antenna mast heights	55.0 m	65.0 m
Feeder cable lengths	70.0 m	75.0 m
Antenna diameter	3.0 m	1.8 m
Transmitter power = 2.0 W	+ 33 dBm	
Diplexer loss	– 2.2 dB	
Feeder loss	– 2.6 dB	
Antenna gain	+ 32.4 dBi	
Non faded free space loss	– 130.3 dB	
Antenna gain		+ 28 dBi
Feeder loss		– 2.6 dB
Diplexer loss		– 2.2 dB

Table 2.3

To obtain the receiver input power, add all the path gains to the transmitter power of + 33 and subtract all the losses:

Non faded input to the receiver	– 46.5 dBm
Minimum operating receiver input	– 81.0 dBm

The fade margin is the difference between the minimum receiver input and the unfaded signal level.

Available fade margin	35.4 dB

Generally, a fade margin of 30 dB is considered acceptable. Hence, this path is slightly better than normal.

2.14 Multipath propagation and diversity operating

2.14.1 Multipath propagation

It is now widely accepted that the earth is not actually flat and so, over a path length of say 30 km, the curvature of the earth will appear as a bulge in the path, and it is actually a curve drawn at a radius from the earth's center. This bulge must be taken into account when planning the height of the antennas at each site.

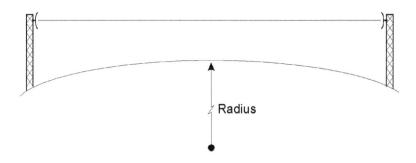

Figure 2.17
Curvature of the earth

In a point-to-point radio system, the path of a radio wave transmitted to a receiving site is planned to pass through free space. It is tempting to think of the beam as a single ray, which aims like an arrow to the receiving antenna. However, this is not the case because when the wave passes through the atmosphere, it encounters many variations in the density of the atmosphere, and these cause the wave to be bent or refracted.

The degree of refraction is not always the same and it changes with atmospheric conditions. It will change quite dramatically in the mornings, when the sun comes over the horizon and begins to warm the upper layers of the atmosphere, and again in the evenings, when it sets and the heat of the earth warms the lower layers. The effect can be very marked in spring and autumn, in dry coastal areas, and is less evident in humid mountainous terrain, where the atmosphere is more diffused.

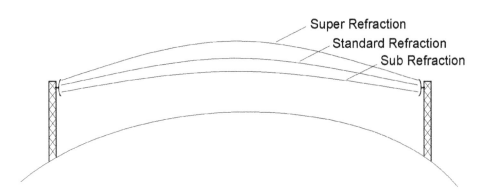

Figure 2.18
Various refractive conditions

So now, there are two curves or radii to be considered – the constant earth radius and the varying curve of refraction of the radio wave. To simplify the task of path planning, these curves can be combined to form a modified or **effective earth radius** curve. If the effective radius is divided by the true earth radius, the ratio K, can be used to define the various atmospheric refraction conditions that need to be considered for various areas of the earth's surface.

$$K = \frac{Effective\ earth\ radius}{True\ earth\ radius}$$

Generally, a value of $K = 4/3$ is regarded as the standard condition. Sub-refraction occurs when K is less than 1 and super-refraction occurs when K is greater than 1. At K equal to one, the radio beam travels in a straight line. Therefore, under conditions of super-refraction the radio wave bulges (upwards) more from the earth's surface and the signal would be able to travel further without grazing the earth.

At K equal to infinity, the beam has the same curvature as the earth, which is referred to as the 'flat earth' situation.

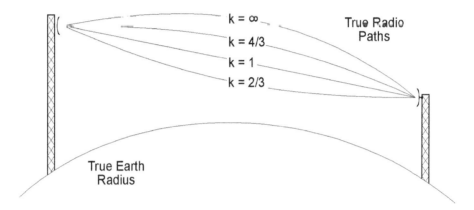

Figure 2.19
Comparison of true radio paths for various K values, with the true earth radius

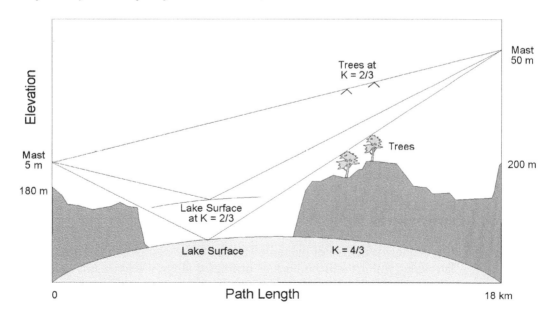

Figure 2.20
A simple path profile

In the Figure 2.20, the radio wave can be drawn as a straight line between the two antennas because the modified earth curvature is drawn with a radius of 4/3, which is the standard condition. However, the radio wave is not a single wave. Because of the uneven density of the atmosphere, the wave is refracted unevenly. While most of the wave follows the same general path, some part of the wave will hit the earth's surface, where it will be reflected onwards, in the same way as a light beam will reflect off a mirror. Some parts of the wave will hit clouds or upper atmosphere formations, where they will also be reflected.

It has been shown that the energy received under free space conditions is the resultant of an infinite number of coherent waves all arriving at the receiver via different paths. All the paths arriving at the receiver antenna dipole, and which are within one half wavelength of the illusory direct path, will be added algebraically. They will contribute their energy to the received signal. The other paths (which may have been more widely refracted or reflected) and which thus arrive from one half to a full wavelength later, will combine to subtract energy from the previously received signal. This adding and subtracting continues with additional wavelength delays. Therefore, the received wave front now begins to look a little like a banana – with many layers of skin arranged like concentric tubes. Thus, the first and third and fifth tubes will all add to the signal and make it stronger whilst the second and fourth and sixth tubes will reduce the signal and sometimes even cancel it out altogether. The so called tubes are really elliptical zones around the direct path line and they are called **Fresnel Zones**, after the man who discovered a similar behavior in light waves. An understanding of the way these zones act is important to the engineer designing a radio path.

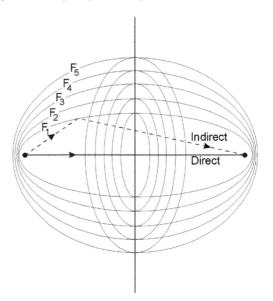

Figure 2.21
Fresnel zone concept

When the centerline of the radio beam is close enough to the surface of the earth and is interfered with, it will be bent or diffracted slightly downwards and the signal will be attenuated. If the diffracting surface is sharp, such as a rocky mountain range, the attenuation will be about 6 dB and if the surface is very smooth, such as a water mass or a desert, the attenuation will be about 20 dB. Therefore, it is important that the centerline of the radio beam clears both the actual surface curve of the earth and the modified K factor curve.

Figure 2.22 is a representation of the way Fresnel zones and diffraction interact, and it will show the way to planning a radio path that avoids the problems described above.

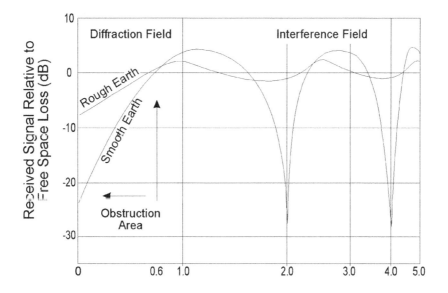

Figure 2.22
Received signal with respect to diffraction and Fresnel zone clearance

There are several interesting points to note in this drawing.

- When the direct beam passes close to the earth's surface, it becomes obstructed, and attenuation begins to rise when the earth's surface or any obstruction exceeds 0.6 of the first Fresnel zone. The attenuation rises slowly for a rough earth profile, and rapidly for a smooth earth profile.
- When the direct beam has a clearance between 0.6 and the first Fresnel zone, the signal is boosted.
- When the direct beam has a clearance around the second and fourth and sixth, etc Fresnel zones, the signal is attenuated. If the clearance is above a smooth surface such as a river or lake, the attenuation can be around 25 dB in the worst case, but if the clearance is over a rough surface such as a craggy mountain, the attenuation will be quite small at about 3 dB.

The vital point to be gained from the above is that paths over a smooth surface must be carefully engineered so that the direct beam lies between 0.6 and the first Fresnel zone. If this is not done, changes in the atmospheric refraction – the K factor – can easily shift the direct beam into the second and fourth Fresnel zones where high-level fading can occur.

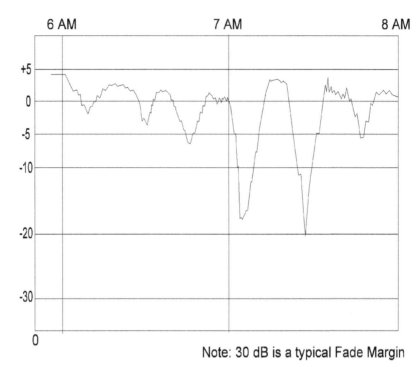

Note: 30 dB is a typical Fade Margin

Figure 2.23
Typical fading characteristics

In Figure 2.23, a graph of received signal strength is shown for a path, which has quite deep attenuation, or path fading. Note that the fades are most severe as the sun is rising, when uneven density of the atmosphere causes severe diffraction.

2.15   Ducting and overshoot

2.15.1   Ducting

Sometimes nature plays strange games with radio signals, which confound the rules laid down for them. In some areas such as the north of Western Australia, the Persian Gulf or the Gulf of Mexico a flat arid land mass lies near a large mass of warm water. Sometimes a layer of air, moist from evaporation from the sea will move over the land mass and will be trapped by the hot air above it. The interface between the two air layers can be quite distinct, and so, a duct of moist air is formed between the earth and the interface. This duct often behaves as a giant waveguide and if a transmitting station is located within the duct, the transmitted signal can travel the length of the duct with very little attenuation.

In the north of Western Australia, 150 MHz transmissions have regularly been received over 400 km from the transmitter and many similar examples have been recorded by radio amateurs who regularly make use of the phenomenon.

Unfortunately, ducting is subject to climatic conditions, is quite critical to operating frequencies, and hence is of little use to the system designer. It can however, cause problems to long link systems serving pipelines and railways, where the tandem paths are sometimes in a straight line.

2.15.2 Overshoot

The problem of overshoot is linked to ducting, but is more a result of poor path engineering than quirks of nature.

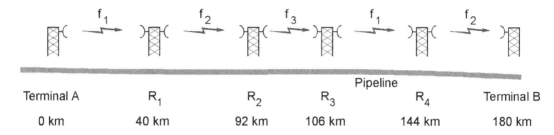

Figure 2.24
A system with overshoot potential

Figure 2.24 represents a logical layout for a system, which might be used to provide communications along a pipeline or a railroad, but it could just as well be a long link between two sites, with two terminal stations and four repeaters. Only one direction of transmission is shown. It is obvious that the transmitter at Repeater 1 must transmit to R_2 on a different frequency to that which it received the signal from Terminal A, and it is normal that the transmitter at R_2 to R_3 will use a third frequency. Where conditions are favorable for ducting, or even if the atmospheric conditions favored super refraction, as discussed in section 2.14, it is possible for the transmitter at terminal A to inject a signal into the receiver at R_4. This is based on the assumption that the receiver at R_4 is tuned to the same frequency as the transmitter at terminal A and that the repeater stations are all in a straight line. It would also be possible for f_2 transmitted from R_1 to get into the receiver at Terminal B. Under favorable conditions, the interfering signal could be almost as strong as the wanted signal and the possibility for severe interference certainly exists.

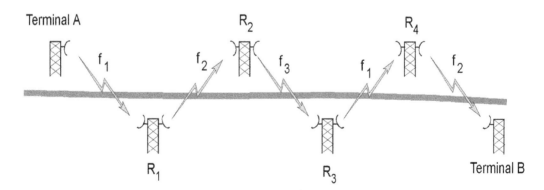

Figure 2.25
A system that avoids overshoot problems

The same type of system is illustrated in Figure 2.25, but this time, the repeaters are arranged with changes of direction in all the paths, and the possibility of overshoot is avoided.

2.16   Duplication of equipment

The reliability of electronic equipment today is extremely high provided the manufacturers' recommendations, regarding operating temperatures and voltages are observed, very few faults should occur. Manufacturers will normally quote mean time between failure (MTBF) figures and if they don't, their equipment should not be used. Given this data, system engineers will be able to estimate the reliability of a complete system, which may be very high, indicating a failure perhaps every two or three years. However, it has to be accepted that equipment will fail and then the time taken to repair the equipment becomes critical, if there is no redundancy in the system.

Most modern equipment is complex and very compact. The very fact that it is so reliable often means that in-house service technicians rarely work on it. Hence, they have little familiarity with it unless the system operator is a large one, with many such pieces of equipment in service. If the system operates in a remote area, the operator will often find it difficult to attract and keep highly skilled technicians, who can repair modern equipment. Realizing these points, manufacturers are building equipment with a degree of internal fault-diagnosis.

A faulty module may be identified on site by an alarm lamp or remotely by means of supervisory systems. In many cases, the faulty module can be removed by an unskilled person and sent to the manufacturer for repairs. Often the manufacturer will be in another country and it would not be unusual for a repaired module to have a turnaround time of one or two months.

If spare modules are held and if they are not tuned to a specific operating frequency, the repaired system may be quickly operating again, perhaps within a few hours.

The system designer must decide if downtime can be tolerated and if so, for how long. If downtime must be minimized, then duplicate equipment must be considered.

Manufacturers generally design their systems to be capable of duplication and there are three levels of protection to be considered.

2.16.1   Cold standby

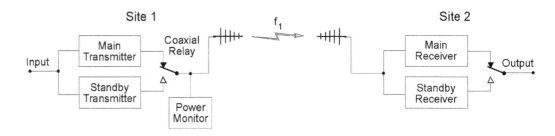

Figure 2.26
Cold standby

In this, the simplest system, the receivers are duplicated and both are fed via a dividing network from the antenna and their outputs are combined or switched to feed the multiplex equipment. At the transmitting end, the output from the multiplex equipment is fed to both transmitters, but it is very difficult to connect the RF outputs together without creating many problems. Hence, one transmitter is fed via a coaxial switch to the antenna whilst the other is turned off. If a failure of the working transmitter is detected, the standby will be turned on and connected to the antenna by the coaxial switch.

There are two problems. First, one cannot be sure the spare transmitter will work as it may not have been used for months and might even have been removed by a thoughtless technician. Secondly, there will be a short interruption to the traffic while the coaxial switch operates and the transmitter comes on line.

2.16.2 Hot standby

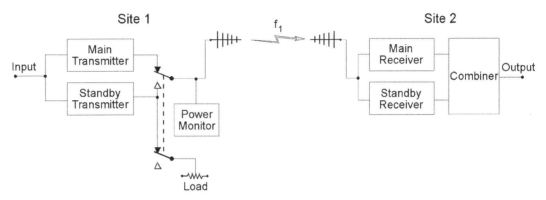

Figure 2.27
Hot standby

This system eliminates the doubt that the standby transmitter may not work, by having it operate all the time and feeding the output via the coaxial switch to a load resistor, to dissipate the power. The switch can operate quickly, so only very short interruptions to traffic will occur, and some manufacturers build a small buffer store into the receivers, so that digital systems will not see any interruption.

There are few problems with hot standby apart from the switching delay, and it is the most favored system.

2.16.3 Parallel operation

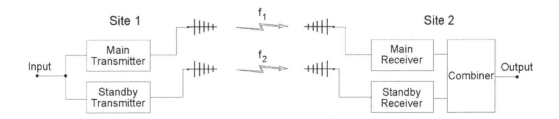

Figure 2.28
Parallel operation

One disadvantage of both hot and cold standby operation is that they both use a single antenna and in some areas, antennas and feeders are vulnerable to damage by birds, wind, salt corrosion, sand erosion or vandalism. Sometimes antennas will need to be removed for repairs or replaced, and if there is only a single antenna in service, long traffic interruptions may occur. Parallel operation uses two antennas at each site, one serving the main transmitter and receiver and the other, the standby pair. If two operating frequency pairs can be obtained, then both main and standby systems can operate in full redundancy, but if only one pair is used, the standby transmitter must feed into a resistive load, as in the hot standby system.

2.16.4   Diversity

Often a radio link will traverse a short path over rocky country and the path engineer can be reasonably certain that deep fade conditions will not be a problem. The main reason for considering the installation of duplicate equipment will then be purely one of equipment reliability and this is discussed in Section 2.17.

Sometimes the radio link must cover a long hop over water and in this situation, despite a well designed path with good fade margins, deep fades may still pose the risk of system degradation or failure. In this case, the designer must consider the benefits of installing additional equipment.

Diversity with radio links is very similar to parallel operation. The idea with diversity is to provide two options for a successful communications path. There are two main diversity methods used.

The first, space diversity, uses two antennas at the receiving end of the link. These are spaced at different heights on the mast. The space between the antennas is specially chosen so that if there is a reflected signal, the two rays arriving at one antenna will be in phase, while the two rays arriving at the other antenna, will be out of phase. Therefore, as the *K* factor changes, there will always be one antenna with two rays in phase. Each antenna is then taken to a separate receiver and the bit error rate monitored from each receiver. The receiver output with the best BER is selected and taken to the output.

Note here that only one transmitter is required. Depending on the trigonometry of the reflection, it may occur only in one direction. Therefore, space diversity may only be required at one site.

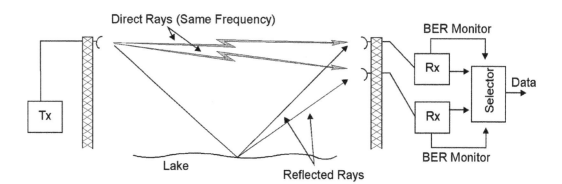

Figure 2.29
Space diversity

The second method is to use frequency diversity. In this case, the same information is transmitted on two different frequencies. Hence, two transmitters are required at the transmitting site and two receivers are required at the receiving site. The theory is that, since different frequencies will have slightly different behavioral characteristics in the atmosphere and when reflecting or refracting off or around objects, one frequency will arrive intact while the other may be corrupted.

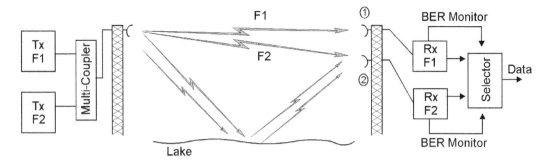

Figure 2.30
Frequency diversity

Note:

A single antenna can be used but generally, space diversity, is implemented in conjunction with frequency diversity.

It is worth noting that space diversity is the preferred option and that generally frequency diversity, is only implemented where very high integrity links are required.

The availability improvement of space diversity implementation, is significantly greater than that gained from frequency diversity. Therefore, it is unusual that it would be economically justifiable to use frequency diversity. If frequency diversity is implemented, then space diversity is provided automatically as part of the system.

2.17 Duplication of routes

There are some situations where some or all of the traffic carried over a radio bearer are so critical that it becomes essential to aim for the very highest levels of availability. Once power supplies, radio equipment and antennas have been duplicated, there is little more that can be done to improve reliability over that particular route. Problems can still occur. In areas prone to cyclones or hurricanes, towers, masts, power lines and even buildings can be brought down. External fires can damage feeder cables, and antennas, and internal fires can completely destroy an installation. In impoverished areas, people may steal the feeder cable, to sell the copper, and there are always the problems of malicious damage by hostile groups, or shooters using antennas for target practice.

There is little that the system designer can do to guard against some of these threats and so alternative routing should be considered where possible and available. It is not possible in a short space to cover all the combinations of risk and alternative routes, so some suggestions are briefly described.

It is possible to run low speed data over single channel VHF or even HF systems. Emergency traffic may be rerouted over such systems, if they can be established.

If public carrier services are available, it may be possible to lease one or two data or voice channels on a permanent or demand basis, to carry essential traffic.

It may be possible to use satellite services on a dial-up basis or even to lease a single data or voice circuit.

3

Satellite systems

3.1   Introduction

In 1945, the famous science fiction writer Arthur C. Clarke, placed an article in the British electronics magazine *Wireless World*, describing how geostationary earth orbit satellites could be placed around the earth, to provide communications to the entire planet. It wasn't until the year 1960 that he was proven correct, and the accolade of being the first person to conceive the idea remains with him.

A geostationary earth orbit (GEO) satellite is one that is positioned in a particular location and distance, above the earth, so that it is orbiting the earth at the same angular rate that the earth is rotating at. Therefore the satellite is positioned over the same spot on the earth at all times. The satellite will be placed directly above the equator at a distance of 35 790 km.

This type of satellite forms the basis for telecommunications links around the world.

A second type of satellite known as the low earth orbit (LEO) is located significantly closer to the earth and moves around the earth in elliptical patterns, generally several times per day.

From the point of view of a telemetry system, it is the services offered by GEO and LEO satellites that would be used. Nevertheless, why would satellite communications be used for telemetry application in the first place?

The following are some of the advantages of satellite communications and why they may be preferred, for a particular telemetry application.

- Satellites provide coverage to large areas of the earth's surface. Virtually 100% of the earth's surface is provided with communications by GEO and LEO satellites
- Normally, distance is not part of the information exchange cost equation. It will cost the same to send a signal over 4000 km as sending the same signal over 1 km
- Modern satellites have capacity to provide wide bandwidth and high data rates between any two points

- The satellite can normally provide communications to any receiver that is within its view
- A satellite is not affected by hills, mountains, deserts, oceans, lakes, urban areas and trees, as are radio and microwave communications systems
- A satellite can provide the same data or voice services to a remote isolated community or industry, as it does to the central business district of a large city
- Satellites can provide sophisticated communications technology to ships, drilling rigs and platforms, located many hundreds of miles from land, and that have normally relied on HF radio for communications
- Satellites can provide voice and data services currently provided to mobile cellular phones to mobile phones/data units any place on the planet

This chapter will examine the types of satellite services that are available, the different service providers and their satellites, the satellite frequency bandplan, the fundamentals of satellite systems, their operation, and then finally, look at the various satellite services available along with their relevance to telemetry.

3.2   Classes of satellite services and their relevant organizations

There are four main classifications of satellites:

- **International**
  Providing global coverage
- **Regional**
  Providing coverage to an area such as Western Europe
- **Domestic**
  Providing coverage to a single country
- **Experimental**
  Providing coverage for research and amateur communications

Any one of the first three can be used for commercial or industrial communications.

The International Telecommunications Union (ITU) is an international organization that represents the interests of over 200 member countries. It is this group that has allocated frequency resources around the planet for satellite communications purposes. Various private or government groups have then amalgamated to put satellite systems into space. It is worth having knowledge of the main satellite systems and supporting organizations.

3.2.1   International services

3.2.1.1   INTELSAT

INTELSAT – The International Telecommunications Satellite Organization – brings television, voice, video, education services, Internet traffic, and private data distribution services to billions of people on every continent through its global satellite system.

Founded in 1964, INTELSAT has provided ground network and mission support for the launch of over 60 of its own satellites. INTELSAT has also made this service available to external customers, and since 1975 the INTELSAT ground network and mission team, has supported the launch of over 60 satellites for those customers. INTELSAT was the first organization to provide global satellite coverage and connectivity, and continues to be the communications provider with the broadest reach and the most comprehensive range of service. Through its role as a commercial cooperative and wholesaler of satellite communications capacity, INTELSAT provides services to billions of people through its signatories in member countries.

INTELSAT's high-power satellite fleet transmits voice, data, and video communications to its customers 24 hours a day, seven days a week. Each day, millions of phone calls, transcontinental travel reservations, financial transactions and database exchanges take place via INTELSAT satellites. In addition, people worldwide are provided international news, entertainment programs and sports events.

After starting operations with the 'Early Bird' satellite in 1965, INTELSAT launched spacecraft to cover the three major ocean regions and established the first global satellite communications system in 1969. Since then INTELSAT has spawned a revolution in international communications capabilities and along the way, has inspired other satellite operators to model themselves after the INTELSAT system.

INTELSAT currently has 20 geosynchronous satellites in operation, providing service to over 200 countries. It currently puts a lot of emphasis on marketing Internet backbone services to carriers and large and small ISPs. It also has a large range of high-speed point-to-point and point-to-multipoint secure dedicated data services. A large range of video services is also available.

INTELSAT provide voice, data, video, Internet, and voice over IP applications to over 400 telecommunications carriers worldwide. For the discerning carrier, their latest satellites provide services that use 'trellis coded modulation in conjunction with 8 PSK and mandatory Reed Solomon outer coding' that enable virtually error free performance.

At the high end of the market, INTELSAT are able to provide STM1 (155 Mbps) dedicated data service to individual businesses or carriers. At the lower end of the market they can provide Internet services directly to the home (through third party service providers).

The Assembly of Parties is comprised of representatives from the governments of all the countries that have signed the INTELSAT Agreement. The Assembly normally meets once every two years to consider global policies and long-term objectives. Signatories are the entities designated by individual governments, to invest in and oversee INTELSAT, in addition to being the main users of the systems. In some countries, the signatory is a private or state-owned company; in others, it is a joint venture or the national government itself. Normally, the Meeting of Signatories takes place once a year to consider broad financial, technical and operational aspects of the system. The Board of Governors, which represents the signatories, meets at least four times a year to decide on all major operational matters of INTELSAT, such as procurement, design, construction, operation and maintenance of the INTELSAT space segment.

For further information about INTELSAT, contact:

INTELSAT Headquarters
3400 International Drive, N.W.
Washington, D.C. 20008-3098 USA
Tel: +1 202 944 6800
Fax: +1 202 944 7898
Web: www.intelsat.com

3.2.1.2   Inmarsat

Inmarsat is a private limited company that provides mobile satellite communications worldwide. Established in 1979 to serve the maritime community, Inmarsat has since evolved to become the only provider of global mobile satellite communications for commercial, distress, and safety applications, at sea, in the air and on land.

On April 15, 1999, Inmarsat, a global mobile satellite communications provider, became the first inter-governmental organization to transition to a private company. Headquartered in London, the company is governed by a 14-member fiduciary Board of Directors.

Inmarsat services

Services supported by the Inmarsat satellite network include direct-dial phone, telex, fax, electronic mail and data connections for maritime applications: flight-deck voice and data, automatic position and status reporting, and direct-dial passenger telephone, fax and data communications for aircraft; and in-vehicle and transportable phone, fax and two-way data communications, position reporting, electronic mail and fleet management for land transport. Inmarsat is used for disaster and emergency communications and by the media for news reporting, from areas where communications would otherwise be difficult or impossible. Systems are also available for temporary or fixed operation in areas beyond the reach of normal communications.

Starting with a user base of 900 ships in the 1980s it now supports links for phone, fax and data communications up to 64 kbps to more than 240 000 ships, vehicles, aircraft and portable terminals. This number is growing rapidly.

Traffic from user terminals passes via the satellites and then down to a Land Earth Station (LES), which acts as a gateway to the terrestrial telecommunications networks. There are approximately 40 Inmarsat LESs, located in some 30 countries.

The satellites

Inmarsat delivers global mobile satellite communications services via its own Inmarsat-2 and Inmarsat-3 satellites. There are a total of four Inmarsat-2 satellites and five Inmarsat-3 satellites. Each of the four Inmarsat-2 spacecraft, which were launched in the early 1990s, has a capacity equivalent to 250 Inmarsat-A voice circuits. Built by prime contractor Lockheed Martin and payload provider Matra Marconi, and launched between 1996 and 1998, the Inmarsat-3s feature spotbeam capability and are each eight times more powerful than an Inmarsat-2.

It requires only three of the Inmarsat-3 satellites to provide coverage to the entire earth's surface except at the poles. The remaining 6 satellites provide backup and extra capacity where it is required.

Due to the increasing demand on the Inmarsat services, it is planned to launch three new satellites in 2004. The purpose of the new satellites will be to support the new Broadband Global Area Network (B-GAN), to deliver Internet and intranet content and solutions, video on demand, video conferencing, fax, email, phone and LAN access at speeds up to 432 kbps almost anywhere in the world. This service will also be compatible with the new 3G cellular services. The new Inmarsat-4 satellites will be 100 times more powerful and have 10 times more capacity than the present Inmarsat-3 satellites.

For further information about Inmarsat, please contact
Inmarsat Customer Care Center
Tel: +44 171 728 1100
Fax: +44 171 728 1110
E-mail: information@inmarsat.org
Web: http://www.inmarsat.org

3.2.2   Regional systems

The INTELSAT, and Inmarsat satellite systems are all current international providers. There are also a number of regional systems, a number of which are noted below.

3.2.2.1   EUTELSAT

More than 30 European countries are signatories to the EUTELSAT organization. There are currently twenty-one satellites in service with plans for more in the future. The satellites service mostly the European continent but also provide coverage to North America and some sections of Asia.

3.2.2.2   INTERSPUTNIK

What was once the Eastern Bloc of countries also has a series of satellites for commercial telecommunications. These are referred to as INTERSPUTNIK and were founded by the Soviet Bloc in 1972. These satellites are not as advanced as the INTELSAT satellites. There are now more than 15 countries that are signatories to this agreement in a similar manner to the INTELSAT agreement.

The services of both satellite systems are now being used by both Eastern and Western countries.

INTERSPUTNIK currently has 4 satellites in operation that cover mostly the European countries. INTERSPUTNIK also now has a very close association with the EUTELSAT organization.

3.2.2.3   ARABSAT

ARABSAT comprises a regional satellite system for the Arab state countries. There are approximately 25 signatories to this organization, who presently have two operational satellites and more planned for the future.

3.2.2.4   Telesat

Telesat is a commercial Canadian company that was the first commercial company in the world to launch a domestic communications satellite in 1972 (as different from Government). Telesat currently have four satellites in orbit and have planned to launch two more. Their market is mainly through North America, with some coverage into South America.

3.2.2.5   PanAmSat

PanAmSat satellite system provides coverage mostly to America and Europe but also provides some coverage into Africa, the Middle East and Asia. The PanAmSat system currently has 23 operational satellites. They claim that they have coverage to 98% of the world's population.

PanAmSat are probably the worlds biggest satellite carrier of entertainment and television broadcast data. They also have a full range of voice, data, Internet and video services.

3.2.2.6   NewSkies satellites

This company owns and operates five satellites around the world. They provide coverage to America, Europe, the Pacific Ocean region, Africa, the Middle East and Asia. They provide video, Internet, high-speed data and multimedia market over their network. They have managed to find themselves niche markets such as providing Spanish and Portuguese language programming throughout Latin America. New Skies plans to launch two new satellites in 2002 and 2003.

3.2.2.7 Skynet

This company owns and operates eight satellites around the world. They claim to provide coverage to 85% of the world's population. Skynet provide a full range of communications services including television broadcasting, direct to the home services, data services, Internet, Video conferencing, voice, distance learning and business television.

3.2.2.8 Others

There are numerous other satellite providers around the world serving regional markets. Some of these are listed below:

- Americon – 18 Satellites
- Astru – 13 satellites
- AsiaSat – 3 satellites
- Nahuelsat – 1 satellite
- Sirius – 3 satellites
- StarOne – 5 satellites
- APT Satellite Holdings

Other regional groups in Asia, South America, and Africa are currently implementing their own regional satellite systems.

3.2.3 Domestic

Many countries have embarked upon establishing their own domestic satellite systems. Approximately 20 countries worldwide have launched their own domestic satellites. This type of telecommunications service is particularly useful to large countries, with remote isolated areas, such as Australia and America. The first country outside of the USA to launch and operate its own domestic satellite system was Indonesia (Palapa, or now called Satelindo). The Australian domestic satellite service was originally launched in 1985, under a government formed enterprise called AUSSAT. In 1992, AUSSAT was sold to OPTUS, as part of the overall agreement, to become the second telecommunications provider in Australia.

Many commercial telecommunications companies buy large quantities of satellite capacity, add cxtra uscr services and facilities, and then re-sell the enhanced services to the public. In the United States of America, a number of commercial conglomerates (normally a posse of state telecommunications carriers) have launched and operate their own commercial satellites.

Of course, in a world all of its own is the plethora of military satellites presently in use around the world.

3.2.4 Low earth orbit (LEO) satellite

As distinct from the GEO, the LEO is a satellite that is designed to orbit the earth at very low (comparative) levels. Sometimes referred to as near-polar orbit satellites, because they have an elliptical orbit that comes close to the poles, they have orbital altitudes that vary from about 500 km to about 1500 km. Kepler's first law dictates that the satellite will have an elliptical path with the earth as one focus. The satellite stays in orbit because the force produced by centripetal acceleration is equal to the gravitational pull of the earth. The period of orbit is determined by the velocity of the satellite. For altitudes of 500 km and 1500 km, the orbital periods are 95 and 116 minutes respectively.

Each satellite will pass twice (in opposite directions) over the same area of the earth's surface each day (as the earth rotates). The satellites can be made to synchronize to the sun, and therefore, pass over the same position on the earth at the same time each day.

Theoretically, the satellites would continue to orbit the earth indefinitely if it were not for the very thin atmosphere that extends into the LEO. This causes drag on the satellite, which will slow it down and correspondingly reduce the altitude of the orbit. Eventually the satellite will slow down so much that it crashes to earth. This could take from several months to one year. Therefore, the satellites continually need to fire on board rockets to keep them in orbit.

LEO satellites are used for mobile telephony, SCADA and telemetry applications, meteorology, surface observation and mapping, research into atmospheric properties and study of land, ocean, and ice regions and properties.

The recently launched Hubble telescope for exploring the outer regions of the universe is aboard a LEO satellite.

There are now two major groups operating large LEO satellite networks for mobile telephony and data applications. The first of these systems is Iridium, built and launched by a consortium led by Motorola. This service was released in the late 1990s and promptly went bankrupt, due mainly to the lack of take up in services. It has subsequently been bought out by another consortium (at a cost significantly less than what it cost to build) and, at the time of writing this book, appears to be operating successfully. The American military have committed themselves to buying significant services from the new consortium.

Iridium is effectively the only provider of true 100% global voice and data coverage of the earth. Iridium consists of 66 LEO satellites that are currently operated and maintained by Boeing. It is designed to provide essential voice communications to remote parts of the planet where normal terrestrial services are not available.

Iridium handsets operate in two modes. When the user is within range of the normal cellular network the handset will operate through this network. When the user moves outside the coverage of cellular networks it will automatically switch to the Satellite mode. Other than a significant increase in call-costs, this switching is transparent to the user.

Iridium sells a number of data services and related hardware to enable the phone to be connected to a computer for data access. They provide essentially two data services. The first is a simple dial up service where the user dials up another computer to make a point-to-point data connection. This operates at a data speed of 2.4 kbps. The second service is for direct connection to the Internet. Using special compression algorithms it is possible to get a data throughput of up to 10 kbps (depending on the type of data being down-loaded).

The second LEO satellite system is provided by a consortium called Globalstar. This system consists of a network of 48 satellites. It does not provide coverage to all of the earth simultaneously. The orbiting of the satellites has been established to concentrate coverage on most of the landmasses with exception of perhaps the African continent.

The functionality of the Globalstar system is fundamentally the same as Iridium. The transmission technology is slightly different. Globalstar uses a CDMA technology developed by Qualcomm for its data transmission. A range of phones and data modems is manufactured by Ericsson, Qualcomm and Telit.

The data service that is provided by Globalstar is based on the serial asynchronous IP protocol using PPP. It provides a maximum data rate of 9.6 kbps with an average throughput of 7.4 kbps. SMS messaging and VPN tunneling for secure connectivity are also available.

3.2.5   Global positioning system

The global positioning system (GPS) is a positioning service that provides position and latitude, longitude to anyone on the planet. Users on the surface of the earth, receive multiple signals from a number of different orbiting satellites, to locate their position on the earth. This service was launched by the American Defense Department as part of the 'Star Wars' initiative during the 1980's. The American government decided to release a limited access code to commercial users around the world. Because it is a defense department initiative there is no guarantee that the service will always be available. If there was a major outbreak of war, it is possible the commercial aspect of the service would be discontinued.

The most commonly used commercial code (there are two) is provided free of charge. This has also caused a lot of concern in the higher circles of American government, as the cost of providing this continuous service is astronomical. There has been discussion in the past of closing the system down to save money. Although this threat is not to be taken too seriously, big potential users should keep it in mind.

There are now 25 satellites orbiting at around 25 000 km from the earth (medium earth orbit – MEOs). This is the final designed full contingent.

The satellites transmit two main codes. The first, the 'P' Code, is used only by the American Military. The code has a variable length of decoding that can run for up to 3 days, making it virtually impossible for anyone else to decode. This can give accuracy of position down to several centimeters.

The second main code used is the 'selective adaptive' (SA) code. This is available to any person with a commercial GPS receiver. It is continually randomly changing (introduced jitter) so that its accuracy can never be predicted. Until recently, the best accuracy that was guaranteed was ±100 meters. The jitter within the SA code has now been reduced and the accuracy of the positioning system has been improved to ±25 meters.

To increase accuracy even further some companies use a technique referred to as differential GPS (DGPS). Here, a GPS signal is received at a known survey point. The difference between the received signal and the known point is calculated, and then transmitted to remote units by radio (or satellite), where the differential signal is then used at that remote site. This technique can produce an accuracy of better than a meter, but the accuracy depends on the number of satellites that can be connected to simultaneously and also on the distance from the known survey point to the remote unit.

3.3   Frequency band allocation for satellites

The overall administration and regulation of all telecommunications standards and practices, for the international community, is undertaken by the ITU.

The ITU undertake all the development of recommendations for technical standards and practices to be used in radio and satellite. A regulatory body, which is under the auspices of ITU, is the International Frequency Regulation Board (IFRB) who undertake to organize and distribute the use of the frequency spectrum, in a fair and diplomatic manner, amongst the international community.

Because satellite, by its very nature, imposes its transmissions on other users in different countries, it must be carefully regulated. Figure 3.1 outlines frequencies that are allocated for satellite communications. The entire allocation is broken down into six separate bands. Each band has a section allocated for *downlinks* and another for *uplinks*. Downlink frequencies are for satellites transmitting to the earth station and the uplink frequencies are for earth stations transmitting to satellites (discussed in the next section).

K_a Band	27.000 to 31.000 GHz uplink
	18.300 to 22.200 GHz downlink
K_u Band	12.750 to 18.100 GHz uplink
	10.700 to 13.250 GHz downlink
X Band	7.900 to 8.400 GHz uplink
	7.250 to 7.750 GHz downlink
C Band	5.725 to 7.075 GHz uplink
	3.400 to 4.800 GHz downlink
S Band	5.925 to 6.055 GHz uplink
	2.535 to 2.655 GHz downlink
L Band	1.550 to 1.600 GHz uplink
	1.500 to 1.550 GHz downlink

Figure 3.1
Satellite frequency band allocation

A major problem that exists with many of these allocations is that they must be shared with terrestrial microwave links. Figure 3.2 illustrates how this can cause interference, to both the satellite and the microwave link.

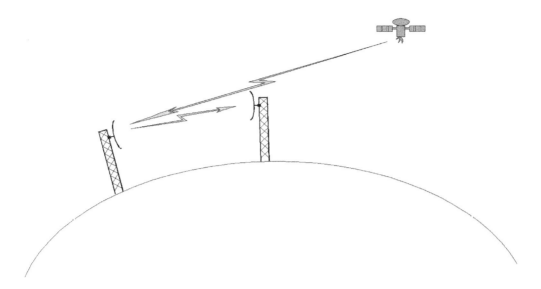

Figure 3.2
Potential interference between satellites and microwave systems

Careful coordination and allocation of frequencies must be carried out by the frequency regulatory bodies to ensure that problems are avoided.

The downlink frequencies are the lower allocation of frequencies for each band; because the RF amplifier is naturally more efficient at producing lower frequencies with consequently lower power consumption, which allows for more transmission capacity at the satellite.

3.3.1 The satellite bands

3.3.1.1 C band

The C band is the most commonly used band for telecommunication satellites. The main reason for this is that, it has a lower natural signal to noise ratio than its surrounding bands (both L & KU). Its main noise component is from the receiver. At frequencies above 10 GHz there is more receiver noise and at frequencies below 1 GHz there is significant background and man-made noise. Interference from terrestrial microwave links is the main problem encountered by C band satellites.

The uplinks and downlinks in the C band are allocated 500 MHz of bandwidth each. Each link is then broken down into 24 wideband channels of 36 MHz. Each wideband channel is then further subdivided, in 800 channels of 45 kHz bandwidth.

It is noted that 24 × 36 MHz = 864 MHz is greater than the 500 MHz allowed. Twenty-four channels are obtained by using the available bandwidth twice. Twelve 36 MHz channels are radiated within the 500 MHz bandwidth using vertically polarized antennas, and twelve more are radiated in the same 500 MHz bandwidth, using horizontally polarized antennas. This, theoretically, gives an effective bandwidth of 1 GHz. With the older analog style satellites, a data capacity of approximately 60 Mbps per 36 MHz channel was possible. The new digital satellites are capable of several hundred Mbps per channel.

3.3.1.2   Ku band

The Ku band of frequencies is fast becoming the more significantly used band. Half of this band is not allocated to terrestrial microwave, which is proving to be a very attractive frequency section for telecommunications users, who wish to avoid possibilities of interference with terrestrial microwave systems.

The Ku band is subject to high signal attenuation due to rain, therefore, care must be taken during the design process to ensure that outages do not occur during heavy rain periods. Calculations must be carried out to determine whether there is sufficient fade margin to overcome rain attenuation and if larger antenna or higher transmit power, may be required. The elevation angle of the antenna also becomes very important, due to high atmospheric absorption at low elevations.

3.3.1.3   L band

The L band of frequencies is very limited in channel bandwidth, but is excellent for use with mobile receivers, such as those used with the LEO satellite networks. Because of the low frequency, there is less atmospheric attenuation and only very small antennas are required to receive a relatively strong signal.

This band is mostly used for maritime communications (Inmarsat) and for LEO mobile communications. Because of the limited bandwidth, it will only carry very slow speed data or single voice channels.

It is easy to place a small antenna on top of the ocean-going vessel and carry out successful communications. Figure 3.3 illustrates how the L band is used to communicate to the vessel, and then how the C band is used to relay this to the shore earth station.

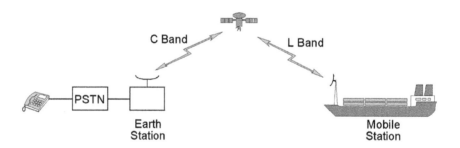

Figure 3.3
L band communications to ocean-going vessels

3.3.1.4   S, X and Ka bands

The S, X and Ka bands are used mainly for government, military and research applications.

There is planned use of the Ka band for commercial applications when the C and Ku bands become overcrowded.

Now that the C and KU bands are very crowded, the Ka band is becoming more popular. Most commercial satellites being launched today will have a proportion of their transponders operating in the Ka band.

3.4 Satellite systems and equipment

3.4.1 Configuration

The earth stations may be classified into four different categories as follows:

- **Main control earth station**
  These are significant installations located at strategic positions on the planet whose main purpose is to provide control and monitoring of the satellites in their GEO orbits. They also provide gateway access to public telecommunications networks.

- **Main access earth station**
  These are normally large installations, which are located at a few major cities within a country, to provide access to public telecommunications networks.

- **Medium capacity remote earth station**
  These provide multiple channels to a user at a particular site, for urban or remote locations. Capacity from these links is normally up to approximately 2 Mbps.

- **Small or mobile earth stations**
  These are small terminals that are located on land-based vehicles, ocean-going vessels or remote fixed locations, providing access for a single voice or data channel.

A fundamental satellite system for point-to-point full duplex communication is illustrated in Figure 3.4.

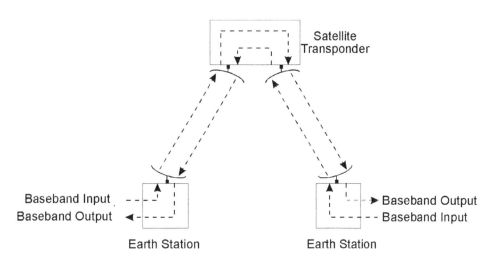

Figure 3.4
Satellite system configured for point-to-point full duplex communications

This type of system will often be used for connection of a dedicated link between locations.

The second configuration is for point-to-multipoint connection. This is illustrated in Figure 3.5.

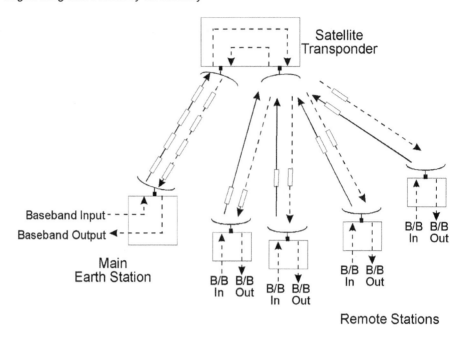

Figure 3.5
Satellite system configured for point-to-multipoint full duplex communications

The main earth station in a country, is normally a large site installation, operated by a main service provider (PTT), located in a few strategically chosen cites, that interfaces directly to the PSTN. The remote stations would belong to any number of different companies, each leasing a dedicated link or paying for time, and accessing the switched telephone or data network. A company would normally dial into or rent a tie line, to the main earth station, to get access to a remote station. The decision to use either dedicated or switched line access to the main earth station is dependent on the access requirements to the remote sites and data speeds.

Packet switched X.25 data access is also available into the main earth stations. Remote terminals will require X.25 modems. Antennas for this service can be as small as a coffee cup.

The medium capacity remote stations are normally referred to as very small aperture terminals (VSATs). A VSAT is a relatively inexpensive earth station normally consisting of a small dish antenna, a relatively high power transmitter amplifier and a high gain, low noise receiver. They are used to communicate a small number of voice circuits (normally 6 to 8) or a number of slow speed data circuits (up to 64 kbps). Higher capacity VSATs are becoming more readily available. VSATs are installed on a concrete pad, aligned and then remain as a fixed installation. A VSAT antenna dish will have a diameter of around 1.5 to 2.5 m.

The main access earth stations can vary, from a very large complex installation by telecommunications carriers that provide a gateway to a major city or state, to a medium size installation set up by a commercial organization, to provide leased services to smaller subscribers. The size of antenna dishes at these sites can vary from diameters of 20 m or more, to dishes of 5 m. The data rate capacities can vary from several hundred Megabits per second, down to 8 or 16 Megabits per second.

There is a range of mobile terminals available. The units will have small flat antennas of 30 cm to 1 m diameter, or long thin antennas about 1 m in height. A small transceiver will be attached to the antenna. The units are designed for use at fixed installations in remote locations, in land-based vehicles or on ocean-going vessels. The units will provide a single voice circuit or single slow speed data circuit (normally up to 9600 baud).

The mobile terminals, normally referred to as mobile satellite service (MSS) terminals, are suitable for remote telemetry applications that require a slow speed data channel. These are the terminals that are used with GEO services.

MSS will be discussed further in section 3.5.

Finally, there are the handheld phone terminals, only slightly bigger than the normal cellular phone. As discussed previously these are used with LEO satellite services.

3.4.2 Multiplexing

For point-to-multipoint satellite systems, there is a need to allocate the frequency bandwidth resource on the satellite to a number of users. There are three main methods used to access the bandwidth resource.

3.4.2.1 FDMA

The first method is referred to as frequency division multiple access (FDMA). This is a concept similar to frequency division multiplexing. Here, separate frequencies are allocated to each terminal station, so the total bandwidth is divided up amongst the potential maximum number of terminals. For a leased dedicated link, a slice of bandwidth is permanently allocated to the terminal. For a switched service, a number of terminals will be allocated a piece of the bandwidth to share and will vie for this bandwidth when it is free for use. The earth terminal will hear everything that it transmits to the satellite because the satellite will retransmit the signal over a wide area, back to the receiving earth station. Therefore, if two terminals are allocated the same slot of frequency bandwidth and try to access that channel at the same time, they will hear the collision interference and stop transmitting. After a random delay, they will try again.

A second method used to allocate switched channel bandwidth is to have a dedicated signaling channel. The main earth station uses this channel to allocate a channel to a terminal.

As noted in the previous section, a C band satellite will have up and down links of 500 MHz bandwidth each, divided into 24 wideband 36 MHz channels, which are divided into 800 single channels each i.e. a total of $800 \times 24 = 19\,200$ channels. In practice, only about 12 000 voice and data channels are available, as capacity is used for other purposes, such as control, signaling, synchronization, etc. These channels are used in two ways. Firstly, a number of channels can be grouped together and sold as a bulk wideband package. This is referred to as multiple channel per carrier (MCPC). A single wideband channel could be broken down and used as follows.

Figure 3.6
A multiple channel per carrier allocation for FDMA systems

It becomes each user's responsibility to carry out the correct wideband modulation using a dedicated carrier. It is these users, who often re-sell the capacity to smaller users (if permitted by the regulatory authorities), by placing a multiplexer in front of the wideband channel and breaking it down into smaller channels.

The other method to utilize the channels on the satellite is referred to as single channel per carrier (SCPC). Here, a single channel is allocated to each user permanently, in the case of a dedicated tie line, and as required in the case of a switched line.

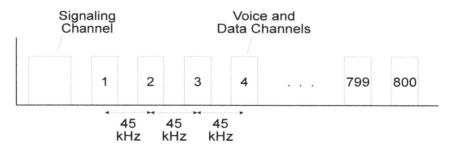

Figure 3.7
A single channel per carrier allocation for FDMA systems

3.4.2.2 TDMA

The second method used to access the channel bandwidth is time division multiple access (TDMA). This is the same concept as time division multiplexing. Here a single large piece of bandwidth is allocated for use by a number of users. Each user has sequential access to the bandwidth for a set short period.

Each remote station is allocated a particular time slot in a frame period, the frame period being equal to the total number of time slots (i.e. the time between each remote station's time slots).

During that time slot the remote station can transmit a short piece of data. This is referred to as a traffic-burst.

An example of a typical frame is shown in Figure 3.8. This frame is for a digital satellite system on one of its wideband channels, operating at 120 Mbps. (Note that some satellite systems can provide channels operating at 4 times this data rate, using more complex modulation techniques.)

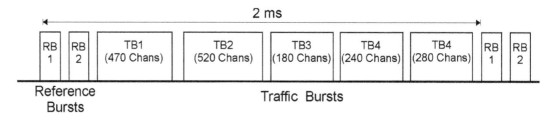

Figure 3.8
An INTELSAT TDMA data frame of 120 Mbit/s wideband channel

Each of the 1690 channels in the frame is allocated a 64 kbps data speed. Therefore there is a 1690×64 k = 108.16 Mbps traffic data speed. The remainder of the 120 Mbps channel is used for control, error correction, and synchronization. The frames are 2 milliseconds in length (that is 1/500 of a second) during which each channel has an allocated time slot to transmit. Therefore 108.16 Mbps/500 = 216 320 traffic bits are transmitted per frame or 128 traffic bits per channel per frame at 64 kbps. There are other reference and synchronization bits included in each frame.

For a satellite that has 24 wideband channels of 36 MHz each (i.e. per transponder) and 1690 channels per wideband channel, there would be 40 000 single 64 kbps channels available. In practice, about 33 000 voice and data channels are available as capacity is used for control, synchronization signaling, etc. This is far better utilization of the bandwidth than using FDMA, which produced a capacity of approximately 12 000 channels for the same bandwidth.

The most difficult thing to overcome in a TDMA satellite system is synchronization. The remote earth stations hear all data that is sent from the satellite, to which they can synchronize. A synchronization pulse is included in the satellite to earth data to provide a suitable reference. Exact synchronization is made difficult because of two factors. Firstly, the satellite drifts in its orbit to a certain degree every day, both in orientation and in direction. Secondly, the distance from the satellite to every point on the earth is not the same. The former problem causes the greater frustration, and sophisticated algorithms are required to overcome the problem.

For fixed services, an earth station is allocated a dedicated slot in a frame. For switched services, when an earth station wishes to make a call, it requests a slot through a signaling channel and is then allocated a slot by the master earth station.

3.4.2.3   CDMA

The third method is referred to as Code Division Multiple Access (CDMA) or sometimes as the *spread spectrum* technique. Here a very wide bandwidth is allocated to the users. A pseudo random code is provided to each station that picks a frequency for a very short period of time, transmits a small number of bits (sometimes just one) and then moves to a new frequency and transmits the next bits in sequence, continuing to hop around the frequency band in an order according to the pseudo random code. This is sometimes referred to as frequency hopping.

The receiver will also have the same pseudo random code and will track the short data transmissions around the frequency band. To other users, who are not part of the system but are on the frequencies that the code selects, the information bits will just appear as background noise. The coded receiver will also pick up noise, but will use error detection/correction techniques to remove them.

This same technique can also be used in the time domain. Here the frequency remains constant but short data bursts at high data rates are randomly spread over a timeframe.

There are a number of advantages with using CDMA. These include:

- Low interference from fixed terrestrial microwave or satellite systems because of the varying frequency
- They require only small low power transceivers and antennas
- High data security
- Immunity to multipath fading because of frequency diversity

The technique is very wasteful of bandwidth because as the wide band channel becomes loaded down, the information signals deteriorate very quickly. The technique is also only useful for slow data rates, up to 9600 baud. Typically, about 800 channels are available on a 36 MHz transponder. For this reason CDMA is mostly used for military applications or for amateur or experimental purposes.

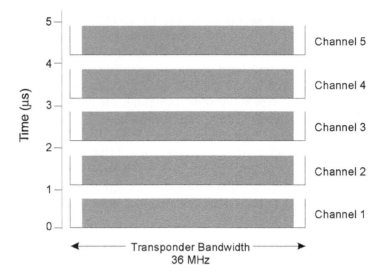

(a) Time division multiple access
  Each channel getting access to the bandwidth for a very short period in a predetermined sequence

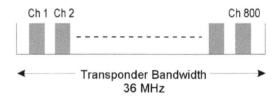

(b) Frequency division multiple access
  Each channel is allocated a set frequency bandwidth within which it can transmit

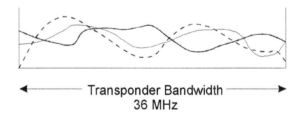

(c) Code division multiple access
  Pseudo random codes are used to spread the channels across the entire bandwidth

Figure 3.9
Comparison of FDMA, TDMA, and CDMA

3.4.2.4 Packet services

Most satellite network providers also offer packet switched data services. These services are generally quick and cheap to implement.

Packet access over satellites is fundamentally a TDMA system, with as many different communication protocols available to access the network, as there are satellite networks. One method is a carrier sense multiple access collision detect (CSMA-CD) type protocol, often referred to as ALOHA (developed at the University of Hawaii). Here, all remote stations attempt to access the channel when they have data to send. Because they can hear the satellite retransmit the data back to the receiving earth station, they can determine if it has been correctly sent. If two remote stations transmit at the same time and a collision occurs, then each station will hear the collision in the form of corrupted data.

Each station then waits for a random period and tries to transmit again. In the case where the transmitting earth stations do not hear the retransmitted signals from the satellite to the receiving earth station, the receiving earth station will send back a *not acknowledge* signal, indicating that the data was corrupted.

Another method involves using a dedicated channel to allow remote stations to reserve capacity, in a similar manner to a signaling channel. When there is free capacity the main earth station signals the remote station to transmit and tells it what time slot to use.

A third method involves passing a small data package, called a *token* between remote stations in a predetermined sequence. Whichever station has the token is allowed to transmit the data. There are limited token hold times; so one station does not dominate the channel.

3.4.3 Modulation techniques

As with the methods of accessing a satellite network discussed in the last section, there are again just as many techniques used to modulate the information signal into the required bandwidth, as there are satellite networks.

The majority of satellites used for commercial telecommunications purposes are moving into purely digital communications mode.

For high speed wide bandwidth links (over 2 Mbps) modulation techniques such as multi-level quadrature phase shift keying (M-QPSK) and multi-level quadrature amplitude modulation (M-QAM) are used. For slower speed links techniques such as minimum shift keying (MSK) are popular.

The efficient implementation, of digital communications over older analog satellites, is proving a particular challenge for satellite operators. Modern digital satellites have onboard channel switching and the ability to direct communications onto different transponders, within the satellite.

3.5 Satellite equipment

This section will examine the various fundamental equipment elements of a satellite system. A complete satellite system is illustrated in Figure 3.10.

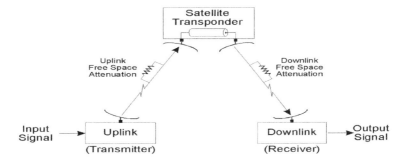

Figure 3.10
Fundamental satellite system

3.5.1   Uplinks

An earth station uplink is the equipment used to modulate the information input signals into a higher frequency carrier signal and then transmitting it to the satellite. The uplink signal includes the information to be retransmitted back to earth by the satellite, plus control information, to operate the satellite's internal functions. Figure 3.11, illustrates the components of an uplink.

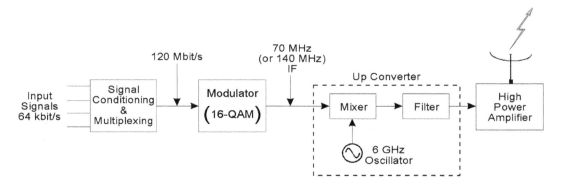

Figure 3.11
Block diagram of satellite uplink

In this example of an uplink, the 64 kbps input signals are multiplexed onto a 120 Mbps composite signal, which is modulated into a 16 part quadrature amplitude modulation (16 QAM) signal and then applied to an FM modulator. The baud rate to the FM modulator would therefore be $120 \div 4 = 30$ M baud.

As each transponder is allowed 36 MHz of bandwidth, the FM signal will have a maximum deviation of ± 18 MHz. The modulator outputs the signal at an Intermediate Frequency (IF) of 70 MHz (note here that the FM modulation index is $\dfrac{18\,\text{MHz}}{30\,\text{MHz}} = 0.6$).

The IF signal is then applied to the UP converter. The output of the mixer will therefore be 6.070 GHz ± 18 MHz and 5.930 GHz ± 18 MHz. The lower sideband is then filtered and the upper sideband output applied to a high power amplifier (HPA) for feeding to the antenna. A typical HPA at a main earth station will operate at 150–200 watts.

3.5.2   Satellite transponder

The satellite transponder receives the signals from an earth station uplink and then retransmits them to earth station downlinks. The receive signals are amplified, shifted down in frequency, amplified again and then transmitted.

A modern satellite will have, as an example, as many as 48 Ku band transponders and 36 C band transponders. With the increased use of the Ka band it is anticipated that the number of transponders that a satellite will carry will increase significantly over the next few years.

Figure 3.12 illustrates a typical satellite transponder.

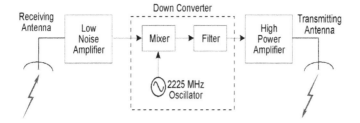

Figure 3.12
Block diagram of a satellite transponder

The 6.070 GHz signal arrives at the receiving antenna and is amplified by the low noise amplifier (LNA). The LNA is a specially designed amplifier that will amplify very low levels of signal that are received at the satellite, without introducing high noise levels. These amplifiers can provide up to 50 dB gain.

The signal is then fed to a down converter, to lower the frequency into the required downlink band of frequencies. The down converter consists of a mixer that is fed with a 2225 MHz oscillator. In the above example the 6.070 GHz is mixed with the 2225 MHz signal to produce 6.070 − 2.225 GHz = 3.845 GHz ±18 MHz and 6.070 + 2.225 GHz = 8.295 GHz ±18 MHz. The upper sideband is filtered out and the lower sideband fed through to the HPA, for amplification to the transmitting antenna. The HPAs in transponders vary from 10–50 watts. Because semiconductors have relatively short lives and are generally unreliable, traveling wave tubes (TWTs), a valve type amplifier, have been used in satellites for many years. Recent advancements in Gallium-arsenide FETs have seen them replacing TWTs.

3.5.3   Downlinks

An earth station downlink is the equipment used to demodulate the signal, sent down from the satellite, back into the information signal that was originally fed into the uplink. The downlink signals also include information from the satellite on the status of its internal functions. Figure 3.13 illustrates the components of a typical downlink.

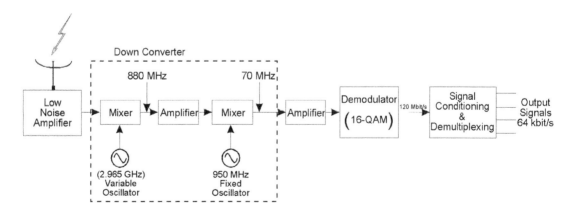

Figure 3.13
The components of a typical downlink

The received signal, which is extremely weak, will be amplified by the LNA by approximately 50 dB. This signal is then applied to the down converter. The down converter consists of a two-stage mixer. The first has a variable oscillator whose frequency is selected to give an 880 MHz output. In the case of the 3.845 GHz received signal, discussed in the last section, the oscillator will be operating at 2.965 GHz (i.e. 3.845 – 2.965 GHz = 880 MHz). This is then applied to a second mixer with a 950 MHz oscillator, which produces the required 70 MHz intermediate frequency.

This is then demodulated to produce the 120 Mbps original digital signal.

3.6   Antennas

The antennas that are used on the satellites and on the uplinks and downlinks are parabolic dishes.

3.7   Link equation

The following is an example of a typical path loss analysis for a satellite system.

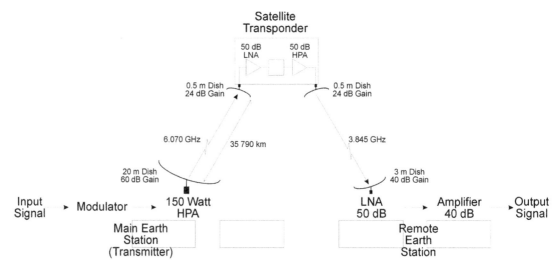

Figure 3.14
Typical parameters for a satellite link system

The free space path loss is calculated from the following formula.

Attenuation = 92.4 + 20 *LogF* GHz + 20 *LogD* km

.·. *at* 6.070 GHz
Attenuation = 199 dB

at 3.845 GHz
Attenuation = 195 dB

If we assume waveguide and connector losses of 6 dB total then total system losses are:

Losses = 199 + 195 + 6 = 400 dB.

The total system gains would be equal to the antenna gains, plus the LNA and HPA gains of the transponder, plus the LNA and amplifier gains of the remote receiver earth stations. Therefore gains are:

$$
\begin{aligned}
\textit{Antenna} &= 60 + 24 + 24 + 40 \\
&= 148 \text{ dB} \\
\textit{Transponder} &= 50 + 50 \\
&= 100 \text{ dB} \\
\textit{Receiver} &= 50 \text{ dB} + 40 \text{ dB} \\
&= 90 \text{ dB} \\
\textit{Therefore total gains} &= 148 + 100 + 90 \\
&= 338 \text{ dB}
\end{aligned}
$$

The main earth station output in dBms:

$$
\textit{Transmitter gain} = 10 \, Log \left(\frac{150}{10^{-3}} \right)
$$

$$
= 52 \text{ dBm}
$$

The link equation then becomes:

$$
\begin{aligned}
\textit{Level} &= 52 \text{ dBm} - 400 \text{ dBm} + 338 \text{ dB} \\
&= -10 \text{ dBm}
\end{aligned}
$$

Therefore, the output signal from the receiver will be -10 dBm.

It is worth noting that the signal level that arrives at the receiver antenna is -140 dBm. This represents a level of 1×10^{-17} watts) i.e. 22 nanovolts. Because of this very small signal level, the receiver dishes must be carefully aligned to the satellite.

3.8   Footprint

The radiation pattern from the transmitting antenna on the satellite will determine the area of coverage it will provide on earth. The direction in which it is pointing and the antenna's beamwidth and gain, will dictate where on the surface of the earth it can be successfully received. The area that it covers is often referred to as the satellite's *footprint*. Figure 3.15 illustrates the footprint for the new L band transponders on the Optus satellites covering Australia.

The numeric values on the concentric circles show the EIRP lines in dBw from the satellite antenna.

Figure 3.15
Footprint for the new optus L band transponders

New techniques of antenna beam-switching and phased arrays allow more control of the directionality and sizing of satellite footprints.

4

Reliability and availability

4.1   Introduction

Long-term operational success, of a telemetry system, is dependant upon two factors. The first is reliability. This is a measure of the quality and system performance, of the equipment itself, over a period of time. Reliability figures will provide an indication, of how well an item of equipment can be expected to perform, for a specified period, under a stated set of operating conditions.

Reliability figures are directly affected by the quality practices used during the manufacturing process, the conditions under which the equipment operated over its lifetime and the level of maintenance undertaken. On the surface, it may appear that the more costly an item, the more reliable it should be, but this is quite often far from the truth (the most expensive cars are not always the most reliable).

The second factor that will affect the success of a telemetry system is the performance of the communications links, between the central control and master site, and between the master site and the RTUs. The performance of these communications links will be measured by the average signal to noise ratio over a period of time, the magnitude of transient noise, the degree of degradation of frequency, phase and amplitude characteristics over a period of time, and what period of time, over a year, the link is available for uninterrupted use. The latter consideration is referred to as link availability.

The other (former) considerations will directly affect the speed of data that can be sent and the number of data bit errors that occur over a period. The measure, of the number of errors that occur over a specified period, is referred to as the bit error rate (BER). When the number of errors that occur on a link becomes so high that the data information cannot be interpreted, then the link is said to be unavailable. Therefore, availability and BER are closely related.

It is important to determine how a communications link is expected to perform, both during the design stage using theoretical and empirical mathematical models and then after installation is complete, using short-term testing and during measurement methods.

This section will examine reliability with relevance to an overall telemetry installation and then will look at availability, with respect to the various types of telemetry communications links that have been discussed in the manual.

4.2 Reliability

4.2.1 Definition of reliability

The measure of reliability is the number of times a piece of equipment fails over a period of time. This is referred to as the **failure rate**. The failure rate, which is of importance, is the number of times an item fails over its life expectancy. Therefore:

$$\text{Failure rate}\,(\lambda) = \frac{\text{Number of failures}}{\text{Total life expectancy}}$$

λ is normally expressed in hours, i.e. *number of failures per hour*

The reciprocal of failure rate (i.e. $\frac{1}{\lambda}$) is referred to as mean time between failures (MTBF). This is the number of hours between failures.

Reliability is of course a game of statistics. If a manufacturer is to publish an MTBF figure, it is fair to assume that he has taken a large sample of his products and run them all through their total life expectancy. This works fine for items such as light bulbs and potato peelers, but it is impossible with such items as radio transmitters, which have life expectancies of 10 to 15 years. By the time an MTBF figure had been determined, the particular model of radio would be obsolete. Therefore, larger samples are tested for shorter periods and the results are extrapolated. There is another, more complex methods of determining MTBF for a piece of equipment.

To explain this method, consider a radio, which is constructed of hundreds of electronic components. A majority of these components will have been available for many years and will have a recorded MTBF provided by the component manufacturers. To calculate the MTBF for the complete radio unit, the MTBF figure for each component is added using parallel and serial probability analysis and the total MTBF figure provided. The former approach is preferred, as it tends to provide a truer indication. The latter approach is often used for calculating MTBF figures, for military equipment. From a project perspective, it is important for the user who is about to procure and install a new telemetry system, to have control over the reliability the final installed system will provide.

To achieve this, attention should be paid to the three following areas:

- Manufacturing process
- Operating environment
- Maintenance procedures

Failure   ssibilities

From a reliability perspective, a telemetry radio system is a complexly engineered system that has many possibilities for failure. These include:

- Failure due to components within products
- Failure due to sub-systems
- Failure due to design flaws
- Failure resulting from software applications
- Failure due to human factors or operating documentation

- Failure due to environmental factors
- Common mode failure whereby redundancy is defeated by factors common to the replicated units

It is not possible to do an analysis that encompasses all these factors, therefore the result of availability/reliability analysis is at best a narrow approximation of the worst factors. Reliability prediction techniques used throughout various industries are mostly confined to the mapping of component failures to system failure and do not address these additional factors. Methodologies are currently evolving to model common mode failures, human factors failures and software failures, but there is no evidence that the models that will emerge will enjoy the level of precision that the existing reliability predictions achieve based on hardware component failures.

4.2.2   Manufacturing

Today more than ever, manufacturers are striving for quality control in their manufacturing process. Manufacturers that are accredited to a national or international quality assurance standard, are required to inspect, test and document equipment at various stages during, and as a final product of, the manufacturing process. If the individual equipment items of a telemetry system are being manufactured locally, it is not unreasonable to request an inspection of the manufacturing facilities. If it is in another city or country this may not be feasible.

During inspection, some items worth checking are:

- Cleanliness of workplace
- Quality of tools
- Regular inspection points being used
- Orderliness of operation
- Documentation at each inspection point
- Marking of components after inspection
- Testing of individual circuit boards
- Quality of test equipment
- Quality of components used
- Inventory procedures for components
- Packaging and storing of finished products
- Attitude of the worker and supervisors to the quality assurance system

If it is not possible to inspect the manufacturing of the various individual equipment items, it is essential that a number of inspections be made during the construction of the telemetry system. The prime system contractor should carry out design, system construction, and testing, locally or in the nearest major city, as this is where the maintenance resources will come from when the system is installed and becomes operational, on site. This ensures that the supplier's personnel, who are to maintain the system, will retain some expertise of the system.

Inspections should be made during several stages of the system design and construction, but most importantly, during the final factory testing. Testing of all parameters of all equipment in the telemetry system should be carried out successfully and documented before the system is accepted as being ready for installation.

4.2.3   Operation

Once a system is installed and operational, it has three stages to its life. The first stage is where there are significant equipment failures shortly after installation, with the number of failures quickly reducing over a short period of time. This first stage is due to design errors or quality-related manufacturing deficiencies.

The second stage is for the majority of its life where the failure rate tends to be low and constant. The third (final) stage is where the system is drawing to the end of its life, and the failure rate increases due to aging and deterioration of equipment. This is illustrated in Figure 4.1.

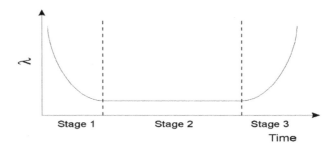

Figure 4.1

During the stage-2 period, the best method to keep the failure rate to a minimum is to ensure that the equipment operates at, or below, its recommended operating parameters, i.e. factors such as output power load and ambient temperature. Other considerations to keep in mind are, to avoid rough handling, corrosive gas environments, moisture, dust, extended operational duty cycles, excessive heat or cold and lack of required maintenance.

4.2.4   Maintenance

All equipment in a telemetry application will require regular maintenance, in particular at the RTU sites. Planned regular maintenance is required to ensure the designed reliability, of the system, is maintained over its lifetime. There should be good quality repairs carried out on faults that do occur and preventative maintenance carried out at planned regular intervals, to assist in finding problems, before they turn into outages.

4.3   Availability

All communications links, used in a telemetry system, should be designed for maximum availability within the budget allocated. This section will examine the availability considerations, and calculations, for the various communications links discussed in this book.

4.3.1   Radio and microwave

To define what the performance of a radio link should be is often a difficult task. The user must decide what level of degradation of the link is acceptable. The preferred answer to the question is no degradation, but radio transmission is statistical in nature, and hence, this is an impossible demand. The user must accept the reality that there is the possibility of outages on the radio link, and must therefore, decide upon an acceptable level of availability.

The major reason for degradation of a radio link is from multipath fading. The signals that arrive at the receiver can come from multiple paths, due to reflections off land or ocean, refraction, ducting, reflections off particles in the atmosphere, and at higher microwave frequencies, fading from rain, fog and other, changing atmospheric conditions.

Other factors that affect the performance achieved include, the time of year (multipathing is more prominent during summer), the noise figure of the receiver (SNR at output compared to SNR at input), the fade margin, the use of single, or diverse radio systems, the link distance and externally produced RF noise.

Link performance is measured as the percentage of time the link is available over a given time period, when compared to a minimum acceptable BER performance.

For example, the performance requirement may be that the link has an availability of 99.9% for any 12-month period, with respect to a maximum BER of 1×10^{-6}.

This means that the system shall not have outages over a 12-month period, that total greater than (365 days × 24 hours/day × 60 minutes/hour × 0.1%) 8 hours, 46 minutes, where an outage is considered to be more than 1 bit in error for every million bits transmitted.

Another example, with less stringent parameters may be a performance requirement with an availability of 99.85%, for the worst performing month of any year, with respect to a maximum BER of 1×10^{-3}.

This means that the system shall not have outages in the worst performing month that total greater than (31 days × 24 hours/day × 60 minutes/hour × 0.15%) 1 hour, 7 minutes, where an outage is considered to be more than 1 bit in error for every 1000 bits sent.

The higher the availability that is demanded from a system, the more the cost involved in achieving this. For example, going from 99.95% availability to 99.995% availability for a microwave link may require using space or frequency diversity in system, which could add 50% to the total cost of the link.

4.3.1.1 Internati nal standard

When measuring BER for high-speed data circuits over microwave links, the international standard requires that the BER be tested over the 64 kbps components, of the high-speed link (often 2 Mbps). Unavailability is then defined as, the periods of time when the BER in each second exceeds $10^{-3,}$ for a period of 10 consecutive seconds or longer, at a data rate of 64 kbps. If the BER exceeds 10^{-3} for only nine consecutive seconds, the link is considered to still be available.

Other terms used for these types of links are severely errored seconds, where the BER exceeds 10^{-3} for one complete second (i.e. a second with more than 64 bits in error in them) and errored seconds, where there are one or more bits in error in a second.

The following table illustrates what availability represents, in terms of system outage time.

AVAILABILITY %	OUTAGE TIME %	OUTAGE TIME PER		
		YEAR	MONTH	DAY
0	100	8760 hours	720 hours	24.0 hours
20	80	7008 hours	584 hours	19.5 hours
40	60	5256 hours	438 hours	14.6 hours
50	50	4380 hours	360 hours	12.0 hours
60	40	3504 hours	292 hours	9.7 hours
70	30	2628 hours	219 hours	7.3 hours
80	20	1752 hours	144 hours	4.8 hours
90	10	876 hours	72 hours	2.4 hours
95	5	438 hours	36 hours	1.2 hours
98	2	175 hours	14 hours	29.0 minutes
99	1	88 hours	7 hours	14.4 minutes
99.9	0.1	8.8 hours	43 minutes	1.44 minutes
99.99	0.01	53 minutes	4.3 minutes	8.6 seconds
99.999	0.001	5.3 minutes	26 seconds	0.86 seconds
99.9999	0.0001	32 seconds	2.6 seconds	0.086 seconds

Table 4.1

This table provides an idea as to what is an acceptable level of availability, knowing how much downtime a particular application will tolerate. The following table gives a more realistic idea, of what may be expected from a link system, for different levels of system importance or integrity. Note though that these are only rough guides, and the eventual availability depends on the system design parameters.

	YEARLY		WORST MONTH	
	AVAILABILITY	BER	AVAILABILITY	BER
Low integrity · Radio link · Microwave link	97% 98%	1×10^{-3} 1×10^{-3}	96% 97%	1×10^{-3} 1×10^{-3}
Average integrity · Radio link · Microwave link	99% 99.9%	1×10^{-3} 1×10^{-3}	98% 99.5%	1×10^{-3} 1×10^{-3}
High integrity · Radio link · Microwave link	99.9% 99.95%	1×10^{-6} 1×10^{-6}	99.5% 99.9%	1×10^{-6} 1×10^{-6}
Very high integrity · Radio link · Microwave link	99.99% 99.999%	1×10^{-6} 1×10^{-6}	99.9% 99.99%	1×10^{-6} 1×10^{-6}

Table 4.2

These figures are for an installed link in the field. If the two ends of the radio, or microwave links, were to be connected together with a piece of coaxial cable, in a noise-free factory environment, then there should only be a residual BER of no greater than 1×10^{-9} from the equipment. There should be no outages and if there were, it should be considered an equipment fault, unless some outside noise source can be identified.

To test the performance of an installed radio or microwave link, BER test equipment is required. For high-speed links of 48 kbps to 2 Mbps, these units are quite expensive. For lower-speed links, of less than 20 kbps, they are significantly cheaper. These units are readily available for hire.

The tests should be carried out for a minimum of 24 hours, with 48 or 72 hours being preferred. The tests should be in loop back mode (i.e. looped from the receiver output to the transmitter input at the far end), for both ends of the link.

There is often uncertainty as to whether to specify the BER at 1×10^{-3} or 1×10^{-6}.

As BER is usually monitored in a system with respect to one-second intervals, then the following applies:

- For 2 Mbps link with a BER of 1×10^{-6} – maximum of 2 bits in error per second
- For 2 Mbps link in the BER of 1×10^{-3} – maximum of 2000 bits in error per second
- For 64 kbps link with a BER of 1×10^{-6} – maximum of 0.064 bits in error per second
- For 64 kbps link with a BER of 1×10^{-3} – maximum of 64 bits in error per second

For the international standard of measuring 64 kbps channels, it is therefore appropriate to use a BER of 1×10^{-3}. It is worth noting that the international standard for BER measurement for 2 Mbps and higher speed links (4, 8, 34) is also 1×10^{-3}. The reason for what appears to be a lenient standard is that they assume that the circuits are for voice, and that the higher data speed bit streams are made up of multiplexed 64 kbps voice channels. If data is being sent at high speeds for industrial applications, where data integrity is vital, then BER standards of 1×10^{-6} should be specified.

4.3.2 Landlines

4.3.2.1 Private cables

For privately owned cables, the only real method of determining the anticipated performance of the link is to run BER test over the cables, for as long a period as is practical (up to 24 hours), during the periods of activity where most RF noise is produced on site. This will give an indication of the state of the line, but will not provide any real indication of what availability can be expected in the long term. The degree of noise on the line will depend mostly on the working environment in which it is laid.

For cables, which are to carry modem signals, it is recommended that approximately 15 dB of signal fade margin be allowed for.

4.3.2.2 Switched analog lines

Telecommunications carriers do not officially provide blanket availability for all their switched analog lines. The author would suggest that the availability on lines in an urban, or suburban area, would be around 99.5%, and for remote locations around 99% (the author was unable to obtain any definite figures from carriers).

The other consideration is the probability of congestion. The system is normally designed so that only one, in every hundred calls made, during the busiest hour of the day, will get an engaged signal. For non-priority telemetry links, these figures should not pose a problem.

4.3.2.3 Leased anal lines

As a rule, a carrier will not provide guaranteed availability figures for their switched or leased lines, but will publish an *availability objective*.

For leased analog services, there should be an expected availability of better than 99.5% for any single year.

4.3.2.4 Dedicated di ital service

An example of the performance objectives for this type of service is given below:

- A longterm BER of 10^{-7}
- 99.85% longterm availability
- 99.5% longterm error free seconds (per annum)
- 99.5% one-minute periods with BER better than 10^{-6}
- 99.95% one-second periods with BER better than 10^{-3}

An error free second is defined as one second in which no errors occur. The link is considered unavailable when 10 consecutive errored seconds occur.

4.3.2.5 Packet switched services

An example of the virtual link availability objective, for a packet switched service over a three-month period, is 99.95% within the packet delay objectives defined for the links (one to three seconds).

An example of the MTBF of a packet switched virtual circuit is 55 hours (remembering the failure of a virtual circuit can be due to hardware failure, software protocol violations or customer operation violations).

4.3.2.6 Switched di ital services

For ISDN services, availability of a circuit should conform to the CCITT recommendations for ISDN switched and semi-permanent lines. Fundamentally this is 99.9% availability, measured on a 64 kbps circuit, where unavailability is periods of time, where the BER in each second, exceeds 10^{-3} for a period of 10 consecutive seconds or longer.

4.3.3 Satellites

The availability of an RF signal from a satellite will depend upon where the earth station is placed, within the satellite footprint. If it is on the outer edges of the footprint, the availability will be less than if it were situated in the inner parts of the footprint. In addition, higher frequencies are more affected by rain and fog attenuation than lower frequencies. Therefore, the signal availability for the Ku band can be reduced in heavy rainfall areas. In general, most satellite links will provide availability of better than 99.9%.

A satellite link is normally a single hop communications link that can cover many thousands of kilometers, and therefore the reliability and availability are relatively high, when compared to a terrestrial microwave link covering the same distance. For example, the cumulative availability of a multiple hop microwave link over 2000 kilometers may be 95%, compared to 99.9% for the link over a satellite.

4.4 SCADA system reliability (or failure) rates

An analysis should be conducted of all equipment supplied, as to what the Mean Time Between Failure (MTBF) is, for each item of equipment, and what is the impact on the system, if an item of equipment fails during operation.

Typical worst-case MTBF, in hours, would be of the order of:

SCADA equipment	MTBF
Operator station (master station)	30 000 hours
Operator display	40 000 hours
Remote terminal units	30 000 hours
Telemetry front ends	120 000 hours

Table 4.3
MTBFs for SCADA systems

It may be necessary to have redundant items of equipment in the following areas:

- An operator station (at the central site) – in this context this includes the computer display/(dedicated) keyboard/archiving system/control software/line printers
- Communication media (either landline or radio links)
- Telemetry front-end equipment that connect to the communications medium at the master station and RTU end
- The system power supply (uninterruptible power supply)
- RTUs in the field

It is important to source industry standard equipment, of which, individual components can be obtained from a variety of different manufacturers. This can ensure that the other variable, mean time to repair (MTTR), is consistent over the entire life of the SCADA system.

4.5 Complete system testing

The final and absolute test is to connect the central telemetry equipment to the communications equipment, and at the RTU end, to connect the communications equipment to the data acquisition equipment, and then to turn the system 'On'. It is often worth specifying in a contract that the system be operational, for a period of time (say 2–4 weeks) without a single outage, before the system is finally accepted. A small portion of the contract value can be kept until this criterion is met.

4.6 Improving reliability

There are three main factors that improve reliability, which are as follows:

- *Complexity*: The fewer component parts and the fewer types of material involved then, in general, the greater is the likelihood of a reliable system.
- *Duplication/replication*: The use of additional, redundant, parts whereby a single failure does not cause the overall system to fail is a frequent method of achieving reliability. This factor is probably the major design feature that can be used to improve the order of reliability that can be obtained. Nevertheless it adds capital cost, weight, maintenance requirements and power consumption.
- *Excess strength*: Deliberate design to withstand stresses higher than are anticipated will reduce failure rates. Small increases in strength for a given anticipated stress result in substantial improvements. This applies equally to mechanical and electrical items. For example, configuring a base-station radio to only transmit at a fraction of its rated RF output power, will significantly improve its reliability.

High reliability, availability, safety and maintainability is achieved during the following activities:

- Design:
  Reduction in complexity
  Duplication to provide fault tolerance
  De-rating of stress factors
  Qualification testing and design review
  Feedback of failure information to provide reliability improvement
- Manufacture:
  Control of materials, tools, methods, changes, environment
  Control of work methods and standards
- Field use:
  Adequate operating and maintenance instructions and training
  Feedback of field failure information
  Replacement and spares strategies (e.g. early replacement of items with a known wear out characteristic)

4.7 Reliability calculations

4. .1 ailure Rate

The measure of reliability is the number of times a piece of equipment fails over a period of time. This is referred to as the 'failure rate'. The failure rate that is of most interest is the number of times an item fails over its life expectancy. Despite this, failure rates for equipment items are normally expressed in per million hours or per billion hours. To obtain the failure rate over the life expectancy, the failure rate is simply normalized with the number of hours for the life expectancy.

$$\text{Failure rate} \left(\lambda \right) = \frac{\text{Number of failures}}{\text{Life expectancy}}$$

The figure for failure rate with respect to 1 billion hours is often referred to as 'FITS' (Failure Intervals). Therefore, 1390 FITS is 1390 failures per billion hours of operation.

$$Failure\ rate\ (\lambda) = \frac{No\ of\ failures}{Billion\ hours} = FITS$$

Where: *No of failures* $= FITS \times E^9 hours$

Failure rate, which has a unit of t^{-1}, can be expressed as the number of failures multiplied by any number to a negative power of 10.

4. .2   Mean time between failure

The reciprocal of the failure rate over the life expectancy of the item is referred to as the mean time between failures (MTBF). Therefore, the basic formula for MTBF is as follows:

$$MTBF = 1/\lambda$$

To obtain the MTBF in years the following formula should be used:

$$Y = \frac{No\ of\ failures\ (\lambda \times E^9)}{Year} = \frac{No\ of\ failures}{10^9\ hours} \times \frac{8760\ hours}{year}$$

MTBF (years) $= 1/Y$

$$= \frac{1}{FITS \times E^{-9} \times 24 \times 365}$$

To convert from MTBF back to FITS the following formula will apply:

$$FITS = \frac{1}{MTBF \times E^{-9} \times 24 \times 365}$$

4. .3   vailability

Availability is expressed as follows:

$$A = \frac{Up\ time}{Total\ time}$$

$$A = \frac{MTBF}{MTBF + MDT}$$

Where MDT is the mean down time of the sub-system or item (also known as the mean time to repair (MTTR).

For this formula to work correctly, MTBF must be converted from years to hours. The formula therefore becomes:

$$A = \frac{MTBF \times 24 \times 365}{MTBF \times 24 \times 365 \times MDT}$$

To convert availability back to MTBF figures the following formula is used:

$$MTBF = \frac{MTD}{((1/A)-1)\times 24\times 365}$$

Unavailability is the inverse of availability.

4. .4   Comments on calculations

Reliability prediction for systems assembled from commercial off-the-shelf equipment modules is often hampered due to lack of published failure rate specification for the various equipment items. This was a problem encountered when calculating system reliability for most projects. Therefore, figures are often used that are published in books or journals as generic figures, or figures are applied that come from similar types of equipment.

There are many complex and expensive software packages available for analysing failure and providing reliability prediction. These require significant time and skill to operate, but because of the speed of the computer, they can provide analysis of a greater range of relative parameters to use within the fault tree analysis (rather than chosen absolute ones) and provide an improved confidence interval for predicting the reliability.

4.8   Qualification of the processes

There are many traps and pitfalls with providing reliability calculations. The following are reasons that the process of calculating and providing reliability figures can be flawed.

- There is often ambiguity in the definition of the reliability requirements. People are often unsure as to what level of reliability they are willing to pay for, and what different levels of reliability will cost.
- There are hidden statistical risks. How true is the reliability data that is provided by suppliers? How correct are the techniques used to calculate the figures? Calculation of reliability uses pure statistical methods, and as they say 'there are lies, there are damned lies, and then there are statistics!'
- There is often inadequate coverage of the requirements. Unlimited time and money could be spent calculating reliability figures. As this is not realistic, there are often short cuts taken and many assumptions made.
- There are often unrealistic requirements to be met. There are often many factors beyond the control of the supplier/contractor that will affect the reliability of the system, over which they have no control. This may reduce the reliability below that specified.
- There are often many unmeasurable quantities. In this case, many assumptions are made, which lessens the believability of the final figures.

5

Infrastructure requirements for master sites and RTUs

5.1   Location selection

In general, the location of sites for radio terminals will be determined by existing masts and buildings, or if a new installation is planned, by proximity, to the site the system will serve. Where a new site is to be developed, the comments below should be considered.

The *approximate* location of a site for a radio with an RTU terminal will be quickly resolved; because it will be in the general area of the plant or equipment it is to serve. The location of a repeater station for a link radio system will be in the general direction of the distant terminal or repeater station. At this stage, the system designer needs to decide the frequency band that will be used for the link. If the lower link frequencies, from say 140 MHz up to 4 GHz will be used, the path lengths will be up to 50 km, providing there are no path anomalies, but if the upper frequencies around 35 GHz are to be used, the path length will probably not exceed 7 km.

At this stage the system designer needs to look carefully at the best available survey maps of the area, so that possible sites can be identified, and the following check points will help to select the best location.

- Is the site flat enough to accommodate the proposed station?
- Is there any road access or can a road be constructed to the site?
- Is there a power line nearby which could supply power to the site? (If it is a high voltage line, the cost of installing a substation should be considered)
- Are there military or security areas nearby which may prevent any development?
- Are there any other radio sites nearby which may cause interference to a new system or which may be interfered with by a new system? (Note that radar transmitters can cause severe interference to other systems)
- Is the area vulnerable to forest fires or flooding and will the access road be subject to flooding?

- Are there any height restrictions imposed by proximity to an airport?
- Is the area subject to forest preservation orders or other green limitations?
- Are there any title claims to the area by native people?
- Are there any plans for, or protection against, future building developments that may block the radio path?

Once all the above points have been cleared and a suitable site located, a preliminary path study must be made to verify that the paths to and from the repeater, are viable, and this will involve the path loss calculations covered in Chapter 3.

If the radio paths involved are satisfactory and all other aspects of the site are cleared, the next step will be to obtain the necessary permits to use the site, and to obtain a right of way for any access road proposed. Once these details have been attended to, detailed site planning may proceed.

5.2 Site works and access

Site access will have to be considered as a major factor in the selection of the site. Sometimes the terrain may provide the designer with a magnificent site and at other times, he may be forced to use a difficult site. In either case it may be expensive, difficult or even impossible, to provide road access.

The radio path is always the unbending criterion in site selection, and if road access is impossible or very difficult, then alternative access must be considered.

If the radio equipment is small and does not require a large antenna-support structure, and if solar or wind power may be used, it may be practical to build and maintain the site with the aid of a helicopter. Whilst they are expensive to use, once the site is established it may only require infrequent visits and the use of a helicopter may be much cheaper than building and maintaining a road. However, it can be safely assumed that if a problem develops at the site it will be at a time when there is no helicopter available!

If a road can be brought to within a short distance of the site but cannot be built up some (last) steep incline, it may be practical to install a funicular. Again, this will be an expensive item and it will require power and regular maintenance; but such systems have been used successfully many times in the past.

In other cases, it may be possible to use human or animal carriers to transport the equipment to a site and to maintain a pathway for maintenance staff. Generally, if a walking path can be established, then so can a track suitable for a four-wheel drive vehicle and this will be much appreciated by the maintenance staff.

All the above benefits must be calculated and the benefits carefully weighed against the advantages of the site. At this stage the designer should look again at the path and consider two further options:

- It may be more practical to build a tall mast or tower at a less favorable site, which has good access, rather than use the preferred site where access is difficult or expensive.
- It may also be more economical to build an additional repeater station and to use two easily accessible sites rather than one difficult site.

5.3 Antenna support structures

Several factors will influence the type of antenna support structure but in some cases, it may not be necessary to use any additional structure, for, if the building housing the equipment is strong enough and there are few antennas, it may be possible to mount them on top of the building. Of course, this will only be possible if the height of the building is sufficient. The use of the building itself offers several advantages such as the cost saving of a separate structure and the ability to use short antenna cable runs.

Assuming that a mast or tower is required, the choice between a self-supporting tower and a guyed mast must be carefully considered.

A self-supporting tower will generally be a heavier structure and this will increase material and transport costs. It will generally be a wider cross-section than a mast, allowing more room for the installation of antennas. It will require a much smaller area of land as it will have no guy anchor points and it will, generally, provide more torsional rigidity; a point, which will be discussed later in this section.

A guyed tower will generally be cheaper to purchase and transport, and where a structure over 30–40 m is required, the cost advantage will tend to increase significantly. Guyed masts require three or four large sets of guy anchor blocks, which will be at a radius of around 0.6 of the mast height, and so a large area of land will be required. Guyed masts are generally more expensive to maintain because it is essential that the guy tensions are checked every year and guy wires will probably need to be replaced after ten years.

The starting point in the design of an antenna support is the height and this will be derived from the path calculations. Remember that the cost of a mast or tower will increase steeply as the height increases and this cost should have been taken into account during the site selection process.

The height of the uppermost antenna will be a starting point. Before setting this as the mast or tower height, the designer should give some thought to the height of the next highest antenna and whether any significant separation between the two is required, which may force the upper antenna to be higher. If there is likely to be expansion of the route in future, it is very useful to attempt to include future antennas in the design as the extension or replacement of such a large structure will be very expensive. Once the final height is set, then all the other antennas should be located on the plan, to ensure that there is adequate separation between them.

It is worth noting here that once a few antennas have been installed on a mast or tower, the number can increase quite substantially! Normally the increase will be because of expansion of the operator's plant or services, but often other companies of government authorities will request permission to install antennas for their own services. In some countries, government authorities can put considerable pressure to have their requests supported.

Another consideration at this point is the need for special painting and mast lighting if the structure is near an aircraft route. Reference to the local aviation authority will provide the answer and both items will need to be specified in the purchase contract, if required. Note that the replacement of lamps used for mast lighting can be an expensive task. The lamps used should be long-life types and sometimes dual-lamp assemblies can be used with photoelectric cell monitoring and switchover facilities, as well as an external alarm facility, if both lamps have failed. Where installed, mast lighting is an essential facility, which should be powered by a UPS or at least an essential supply.

The next factors to be considered are wind and ice loading and these are local considerations. In regions subject to icing conditions, the weight of ice can add very substantially to the load on the structure and it must be able to carry this additional load. Mast and tower manufacturers should have the experience to design suitable structures, but the designer must be sure to specify the location, so that the manufacturer will know that he must make allowances for icing conditions. Similarly, thought must be given to the antennas themselves, as performance may be degraded by ice and the antennas may be damaged or destroyed by a large buildup of ice. Where required, manufacturers can provide heating elements and radomes to de-ice and protect the antennas. Naturally, the heating elements must be powered from an essential supply.

Wind loading must be considered in almost all locations. Again, the tower or mast manufacturer will be able to design a structure, which can support itself under the worst-case conditions for the area. The information required can generally be obtained from the meteorological authority, but the local topography of the site must also be considered because valleys can act as wind tunnels and considerably increase wind forces. In many countries, the standards association will publish codes for the design of structures and these should include variations for local terrain.

To give an example of the increasing risk factors in cyclone (hurricane) prone areas, the following chart shows wind load multiplier factors, which are applied for various climatic zones.

For normal risk areas	× by 1.0
For moderate risk areas	× by 1.5
For cyclone areas	× by 2.3
For severe cyclone areas	× by 3.3

There are other multipliers, which cover the terrain, building heights and proximity to other buildings.

It is stressed that the subject of wind loading on towers and buildings is a complex one. The design of suitable structures is covered in depth in Australian Standard AS 1170.2.

Another factor, which may need to be taken into account when a mast or tower is specified, relates to torsional stability. This defines the amount the structure will twist under high wind conditions. To illustrate the point, a large diameter parabolic antenna with a gain of around 44 dB will have a beamwidth to the –3 dB points of 1 degree and if under severe wind conditions, the support structure allowed the antenna to swing two or three degrees off line, a severe loss of received signal would occur. Under unfavorable fading conditions, the path could drop out and cause an interruption to traffic.

Secondly, in areas where very severe winds are expected, the mast or tower must obviously be specified to withstand the wind load. It may be prudent to use antennas of a lower wind specification especially for services that are not regarded as essential. If a very severe wind force is experienced, some of the antennas will probably be blown away and this will reduce the wind load on the support structure, perhaps saving it from destruction. If spare antennas are held on site, it will be a simple matter to replace them whereas it might take weeks or months to replace a mast or tower.

5.4   Lightning protection

Many parts of the world are subject to electrical storms and it is hard to imagine a better way to attract lightning than to put a metal structure on top of a high mountain! Therefore, it is generally safe to assume that lightning will be a problem, but in most cases, it can be safely grounded without fear of damage to the installation. If proper protection is not installed it is virtually certain that at some time, massive damage will be caused by a lightning strike.

To understand the basic methods of lightning protection, it is first necessary to understand the way in which it causes damage. Lightning is caused by a huge buildup of static electricity charge in clouds in the atmosphere. Eventually the potential difference between the charged clouds and the earth becomes so high that the air dielectric is broken down and a spark will leap across the air gap, which will then become ionized.

Once this occurs, the air becomes a luminous conductive plasma reaching a temperature around 40 000°C and a very large current will flow. The current will have a rise time of about 2 microseconds, duration of around 40 microseconds, currents of 15 000 amps are typical, and this figure is often exceeded. Now, the tower or mast, being something like a long piece of wire, will have a certain inductance and when lightning strikes it, a voltage will be induced in the inductor, which is in fact the tower. For a 100 m structure, this voltage may easily reach 250 kV between the top and the base. This voltage will also be induced into coaxial feeder cables and power cables for mast lighting and can cause enormous damage to equipment.

The topic of lightning protection is a broad and important one and cannot be ignored by the system designer. There are some specialist companies who manufacture protection equipment and who often publish papers or guides on the topic. These companies should be consulted on the problems of a particular installation.

A basic list of the elements of a protection system is given below.

- A lightning rod should extend above the top of the tower or mast. This will provide a cone of protection about the structure within an arc of around 15° from the top.
- All antennas should be installed within that protective cone. Sometimes the rod will have radial spikes at the top and some types have used radioactive charges to ionize the top of the protecting rod. This increases the attraction of lightning to the rod and away from the structure or the antennas.
- The lightning rod will be connected to a copper grounding cable or strap, to conduct the current directly to the ground. Sometimes the cable will be a type of large coaxial cable, which conducts the current through the inner conductor and protects the structure and other cables from spurious discharges from the inner conductor.
- The lightning conductor must be grounded at the base of the structure because the charge will jump from a bend rather than go around it.
- The ground connection must provide a low resistance bond to the surrounding earth. It may be necessary to improve this connection by means of conductive soils or a radial mat of metal strips. The jointing of the earthing system is important. Junctions of unlike metals should be avoided (or sealed) to prevent ingress of moisture, which will quickly setup a high resistance film between the different metals.
- Where a guyed mast is installed, the guy anchors must be well earthed and heavy conductors must be used to connect the guy wires to the earth connection. If this is not done, lightning could easily shatter the concrete anchor blocks resulting in the remaining guy wires pulling the mast down. It is good practice to extend the base earth mat to link up with the earthing system of the guy anchors.

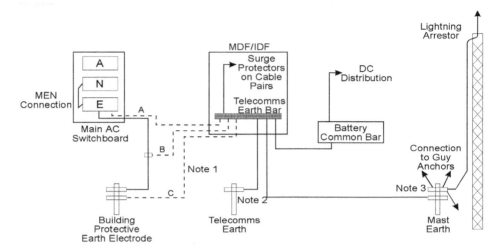

Notes:

1. Telecomms earth connected to AC earth by method A, B or C, but only one connection.
2. Telecomms earth may not be required.
3. Coaxial cables and waveguides should be grounded at the bottom of the vertical run.

Figure 5.1
Communications site – typical earthing plan

- Where a site is powered by means of an overhead electric route, the cable will also be prone to lightning strikes. In order to prevent the lightning energy from damaging communications equipment, the power line should be connected to a step down or isolating transformer located a little distance from the building. The frame of this transformer should be connected by a heavy cable to an earth point. The transformer and power line should be protected by fuses or circuit breakers, in accordance with local electrical practice. Many specialist companies also produce a wide range of secondary protection against lightning induced surges, which are designed to be connected across the low voltage mains supplies and it may well be worth installing this additional protection, in high risk situations. The 110 or 240 volt ac distribution board is connected to a main protective electrical earth electrode.
- Should there be any incoming telephone cables entering the building, they must be terminated onto a jumpered MDF, which is fitted with surge protection devices and the telecommunications earth bar of the MDF should be connected to the main electrical earth point.
- An earth cable should be run around the periphery of the building and where possible, all the steel support members of the building and the roof should be bonded to this cable. This earth is bonded to the main electrical earth.
- A battery common bar should be installed within the building and this should also be connected via a separate cable to the MDF earth bar. To this earth bar should be connected all the internal earth points, such as rack earths, distribution earths and the ground side of the main battery system.

The principle of this rather elaborate installation, derives from the fact that we cannot avoid lightning and nor can we even attempt to dissipate the voltages and currents involved. If, however, all the metallic elements of an installation are securely bonded together, when lightning strikes, the entire installation will rise in potential, to perhaps several hundred thousand volts above ground potential, in unison and decay again. This is subject to the use of low resistance cables, with an efficient connection to the ground, and lightning strikes may include, the power line, the antenna support or the entire building. But the critical point is that there will be only minor differences in potential between say the power supply, the transmitters and the antennas and even persons who may be in the building and the chance of serious damage will be minimized.

5.4.1 Levels of lightning protection

Lightning protection is not an absolute. Protection implies safety and safety is not an absolute. There are only levels of safety and as such, there are only levels of lightning protection. A city manager once asked an electrical engineer, 'Why can't you protect your equipment from a little bit of lightning?' This is tantamount to asking, 'Why can't you protect your equipment from a "little" atomic bomb?' If God looks down and decides that he doesn't like your equipment then it is out of here! The worst-case scenario is that there is going to be a big hole where your cabinet used to be. It is impossible to protect equipment completely from lightning. There are only levels of protection.

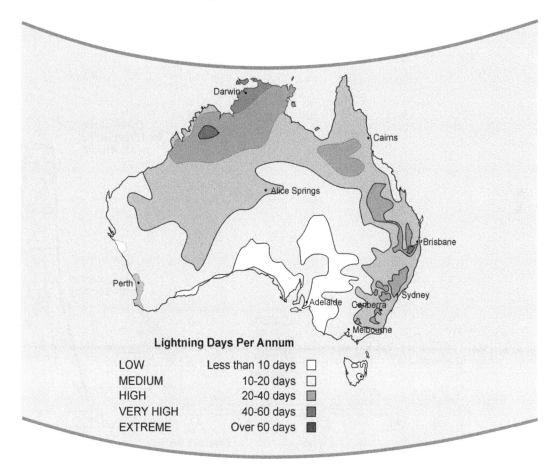

Lightning Days Per Annum

LOW	Less than 10 days	☐
MEDIUM	10-20 days	☐
HIGH	20-40 days	▨
VERY HIGH	40-60 days	▩
EXTREME	Over 60 days	■

Figure 5.2
Lightning days per annum

5.4.2   Separating equipment and lightning

Digital equipment is very sensitive to both high voltages and currents from outside sources. Levels of lightning protection are based on keeping these high voltages and currents away from the equipment. The optimum solution for any protection is to keep the problem away. This may sound simplistic but all protection theory revolves around distance. From boxing to fire protection, distance is the ultimate protection.

Distancing the equipment from the lightning can be accomplished in one of two ways.

1. Designing systems that encourage the lightning to dissipate elsewhere.
2. Move the high voltages and currents coming into the equipment to ground, before they can do any damage.

5.4.3   Dissipating the lightning elsewhere

Since the days of Benjamin Franklin there has been a debate on whether the lightning rod attracts lightning, or whether it only dissipates lightning when it is going to strike anyway. Encouraging the lightning to dissipate elsewhere is done by installing lightning rods, some distance from the equipment. These lightning rods are usually very tall and strategically placed on the down lightning side (if possible) of the property. Heights can be as high as 20 or 30 meters. The rod is held in place by use of guide wires and together these act like a lightning attraction device. The rod is well grounded and often a site will employ multiple rods separated by 20 to 50 meters.

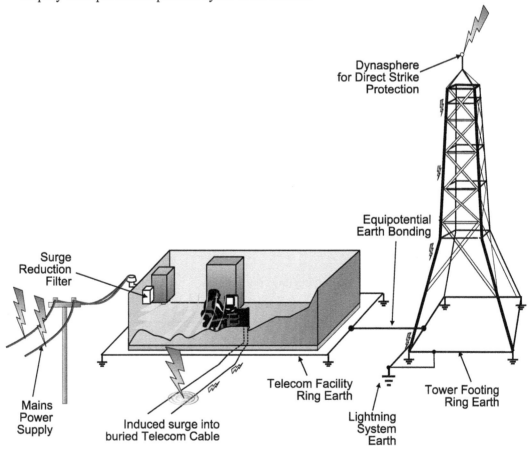

Figure 5.3
Typical earthing system

5.4.4 Dissipation of high voltages or currents

Dissipation of high voltages and or currents to ground, once they come into the equipment, affords the least protection. If the voltage and or current are high enough then no protection is going to help. In the design of digital systems, it is common to connect lightning protection devices to the equipment with the hope of saving the equipment from a strike.

5.4.4.1 i h current r tecti n

High current protection is thought of as the easiest. This is done by putting in a fuse. Due to the fast nature of lightning, the fuse may not have time to blow before the damage has been done. Nevertheless, it is a level of protection. A fast blow fuse or breaker is best.

5.4.4.2 i h v lta e r tecti n

High voltage protection is done by placing devices on the input that short the high voltage, coming in to the equipment, to ground. There are three basic types of devices that can short high voltages to ground:

1. Gas discharge tubes or GDTs
2. Metal oxide varistors or MOVs
3. Transorbs

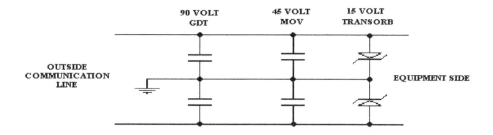

Figure 5.4
Example of a lightning protection circuit

These devices have three basic attributes, current, speed, and voltage level to turn on. All of the devices act like an open circuit under normal conditions. However, when large voltages are present in the equipment, they turn on, and short the offending voltage to ground. Placement of these devices is very important. Usually the higher voltage devices are placed closest to the field and the lower voltage devices placed nearest to the equipment.

5.4.5 Conclusion

When designing a system to protect equipment from lightning the best method is to use all of the different types of devices available for lightning protection. Just as a fighter goes into the ring with all of the tools and skills he can muster, lightning protection should include all economical means possible to protect the equipment from damage. Some type of method should be employed to move the lightning away from the equipment, and if a strike does happen, then devices should be employed to dissipate the voltage and current to ground.

5.5   Equipment shelters and temperature management

The type of building, or shelter, required to house communications equipment, will range from a pole mounted metal case to house a simple solar powered telemetry system, to a large concrete blast-proof structure to house a major communications system in a hazardous or cyclone (hurricane) prone area. Obviously, the equipment it is to house and the location will determine the type of structure. The notes below will serve as a guide to some of the more important points to be considered.

5.5.1   Temperature management

In many areas of the world, the high ambient temperatures in summer will require that some consideration be given to this aspect of the design. Generally, modern equipment is rated to operate at temperatures up to 45–55°C. Equipment that is below this range should be viewed with suspicion, unless it is to be operated in very cool areas.

Where ac power is available, the building will often be air-conditioned and this is certainly a very satisfactory solution, if the system is reliable. If an unduplicated air conditioner fails and there is no supplementary ventilation, the temperature in a room can quickly rise to high levels and sometimes in humid areas severe condensation can occur.

It is worth noting that in remote areas, where domestic type air conditioning units are used, theft can be a problem as the units can often be removed from the outside of the building. At one location known to the author, the unit was stolen several times so a microswitch was installed to signal its removal. This was connected into a supervisory system reporting back to the local telephone exchange. It was then a simple matter of alerting the local police who would await the thieves at the end of the single access track.

Where a site is without adequate ac supplies, careful thought must be given to the building design so that temperature can be controlled.

An equipment room will have a heat load increasing above that of the ambient temperature, when it is inadequately shielded from the direct sun and also when heat-producing equipment is operated inside. These factors must be controlled.

It is relatively easy to provide effective shading and this is often in the form of a metal sunshade that is insulated underneath to prevent re-radiation of heat from the shade itself. The walls of the building can be similarly protected with sunshade walls made of corrugated sheeting or bricks etc. Some smaller equipment has been installed inside large cement pipes, which were buried underground, to take advantage of the coolness of the earth and this system has many advantages in suitable terrain. It offers good security against vandalism, but there may be problems if there is any chance of water entering the installation and at times, snakes and other creatures have found a way in. Adequate ventilation during visits by maintenance staff is of prime importance, as foul air can accumulate in the chamber.

The other problem to be considered is the heat developed by the equipment itself. Fortunately, modern equipment is very efficient and uses little power, but in larger installations, some form of venting must be provided. Sometimes the warmer air rising from the equipment is channeled into a vent, where it can be vented to the outside air, thus forming a thermo-siphon, which can be used to draw in cooler air from underneath the building. Often wind powered rotary turbines can be very effective in sucking hot air out of a building but are reliant on some wind, which is often not present on very hot days. At solar powered sites, hot days usually mean clear skies and there may be excess solar power available above the requirements of the battery system. This may be used to operate exhaust fans to improve air circulation.

One important, and often neglected, aspect of temperature management concerns battery charging. At sites where batteries are charged by solar or wind power or by non-continuous diesel power, the temperature of the battery must be carefully considered. Many modern lead acid cells are rated for operation up to 55°C. If the cells are heavily charged when the ambient temperature is already high, this limit may easily be exceeded. Generally, solar power supplies will not charge at high currents in relation to the battery size and the problem will not be significant. However, with diesel power the tendency is to run the generating plant at high charge rates for a short time. Where this is the case, the plant should be restricted to running at night, during hot weather months.

When a remote site building is to be purchased, the system designer must quote the maximum expected heat load their equipment will generate and the maximum increase in temperature, above ambient, that must be maintained. For example, if the maximum operating temperature for a system were 48°C and the midsummer temperature at the site could reach 43°C, then the heat load temperature rise must not exceed 5°C.

In general, a heat load temperature rise of about 10°C is readily achieved by good design methods and without forced ventilation or cooling. But in many hot and dry arid regions it is not uncommon to have ambient temperatures of around 46°C and under these conditions an equipment operating temperature of 55°C, with a heat rise of 10°C, is barely adequate.

In other areas, minimum temperatures may be the problem and again batteries need to be carefully considered, for whilst most electronic equipment will operate at –20°C or below, lead acid cells become quite arthritic, typically yielding only 61% of their rated capacity at low discharge rates, and as little as 38% at high discharge rates. This is a very important consideration for engine starting batteries.

5.5.2 Other aspects of building design

There are several other factors to be considered when specifying a communications building. Most such as security, type of construction and termite and rodent protection will be determined by local conditions.

The following may serve as useful checkpoints.

- Ensure doorways and passages are large enough to allow bulky equipment to be brought in or out
- Consider the provision of automatic and/or manual fire protection systems
- Do not install windows unless necessary. (They are security risks and detract from the thermal efficiency of the building)
- Connect all doors to a remote supervisory system so that access to the building can be monitored
- Install a fresh water tank
- Provide adequate first aid facilities, especially where battery systems are installed
- Provide sleeping and meal preparation facilities if staff may be required to overnight at the site
- Provide adequate emergency lighting. (At mains or diesel powered sites, provide some dc lighting operating off the battery bank in the event that all ac power fails)

- In some areas, it may be practical to install a telephone connected to the system controller outside the building. (It may provide a means of summoning help for a traveler in need of help and may prevent them from trying to gain access to the building)

5.6 Power supplies

Power supplies for electronic equipment, particularly in remote locations, are the foundation stones on which a reliable communications system is built. All too often the power supply suffers from a lack of design and maintenance, by those who do not understand the problems involved, and who are, perhaps, more interested in the more technically interesting equipment at the site. The result is usually an unreliable system and if the site is remote, many expensive repairs may be necessary.

The choice of primary power supply type will depend on many factors, and the comments below, will guide the system designer in making the right choice.

5. .1 dc supply and batteries

Most communications equipment operates from 24 or 48 volt supplies whilst some, which derive from a mobile radio background, use +12 volt supplies. It is common to provide a storage battery supply as the direct power source and to charge this battery from the primary power supply. There are many advantages in doing this. Batteries are certainly the least complex power supply and if they are properly installed and maintained, there is almost no possibility of an interruption.

5.6.1.1 Ty es st ra e battery cells

Nickel cadmium

These cells are characterized by very long life, 10 to 15 years being common, and by a high initial cost. The normal cell voltage is 1.5 volts and this decreases, as the cell is discharged, to about 1.1 volts. This results in a rather wide voltage band over a charge/discharge cycle but modern equipment is usually untroubled by this. Nickel cadmium cells are very robust and well-suited to applications where there are high vibration levels, but they appear to be more subject to random failure than lead acid cells. Cadmium is now recognized as a dangerous heavy metal and most countries have strict regulations regarding the disposal of worn out or surplus batteries. Most manufacturers will require information about their installation and will offer a recycling service, which may be included in the initial cost of the battery.

Nickel cadmium batteries are easily damaged by heat and must be kept at low ambient temperatures, particularly when charging, as the internal temperature of the battery can get very high quickly. Internal temperature above 55°C can permanently damage the ni-cads. Care must be taken when rapid charging ni-cads not to continue past the fully charged state. Long-term trickle charging on the capacity/10 rate is acceptable, provided the ambient temperature stays around 20°C.

Lead acid

There has been a great deal of development in the manufacture of lead acid batteries since their humble origin as car batteries. Different types have been produced for specific applications and many of these are relevant to communications sites. The cells have a nominal voltage of 2.0 volts and this remains fairly constant over the charge/discharge cycle. The cells are heavy and older types, with a liquid acid electrolyte, were easily damaged by rough handling.

Sometimes cases cracked resulting in serious spillage of acid. The early use of lead acid cells for communications was almost always, in telephone exchanges, where they spent most of their life in a float charge state, absorbing current peaks and providing occasional short duration emergency power, in the event of failure of the primary supply. These cells were quickly damaged if left discharged for extended periods and were not well suited to repeated charge/discharge cycles. Under these conditions, they required high maintenance and sometimes had a service life of only one or two years. Naturally, replacement costs in remote areas were often very high.

With the development of solar power and low power consumption electronic equipment, a great deal of research has produced new cells well suited to most applications. These modern cells tend to be fully sealed, apart from a safety pressure vent. This means that they can be safely transported and can be mounted sideways in a cube, to conserve space. The electrolyte is contained within a saturated fiberglass mat, which replaces the old separator in a wet cell, and when hydrogen is produced as a result of the charging of the cell, it is recombined within the cell to replace the lost electrolyte. These so called recombination cells need very little maintenance and the stacking ability is an important feature. On one hand, the ampere hour capacity of an existing battery room might be quadrupled, if necessary, by replacing old type wet cells with recombination types or alternatively, only a small battery room may be required, which may be an important cost saving at a remote site. Sometimes no separate battery room will be required.

Whilst recombination cells offer many advantages, they do not, yet, have a long history of reliable operation and the system designer should be certain that they will meet system needs. There are many manufacturers offering so-called *maintenance free* batteries and some of these, particularly in the automotive field, have been very unsuccessful. On the other hand, a leading German manufacturer has been building recombination cells for several years and they have been installed in locations such as the control tower at Frankfurt Airport.

The traditional wet cell will probably all but disappear in the future. Nevertheless, it will not be in the near future, as their continuing development makes them a very reliable choice. This is particularly so for solar powered sites, where they operate under charge/discharge conditions and where the recombination cell has yet to be pushed to its limits.

5.6.1.2 Installation and operation – battery systems

In many countries, the standards authority publishes codes of practice for battery installation and these are often mandatory. The more important aspects of installations are noted below.

- Battery rooms or cabinets should always be vented to the outside air and not into an air-conditioning system
- Batteries should be mounted on strong stands and where wet cells are used, drip-trays are necessary to cope with the possibility of a leaking cell
- Where wet cells are used, some first aid facilities should be provided (a minimum requirement is an eyewash bath. Sometimes showers will be required)
- Battery banks should always have a main battery fuse at the battery bank (it is quite pointless to place this fuse at the end of a cable from the battery as the

cable will be unprotected and damage to the cable might easily cause an explosion)

- Where possible, cell terminals and interconnecting bars should be insulated. (Sometimes metal objects may be dropped onto a battery and an explosion could result. Many UPS systems use battery banks operating at 110 or 220 volts and inexperienced persons often fail to appreciate that not all batteries are 12 volt types. Very severe injuries can result from contact with these high voltage banks. Warning notices should be displayed and access restricted to these installations.)

- Wet battery systems require regular maintenance. Electrolyte levels must be checked and replenished regularly and specific gravity readings taken to ensure that all cells are in an equal state of charge. If some cells become undercharged, this imbalance will increase with successive charge/discharge cycles unless an equalizing charge is given. This is caused by stratification of the electrolyte when the acid tends to settle towards the base of the cell. This creates a situation where the electrolyte at the bottom is very strong whilst that at the top is under-strength and so the plates of the cell are unevenly worked. Stratification is overcome by an equalizing charge, which continues the charge cycle beyond the normal cutoff point so that bubbles of hydrogen begin to form. As the bubbles rise up through the electrolyte, they effectively stir it up and thus restore a balanced mixture of acid and water.

- The battery charging system must be able to supply the full normal equipment load as well as to recharge the battery following a discharge cycle. It must also maintain battery float voltages accurately because under-charging will significantly reduce battery capacity and over-charging will shorten the life of the battery. The battery manufacturers' recommendations for float and equalizing voltages should be closely followed.

5. .2   Mains power supplies

Where a communications site is in a town, a city, or an industrial plant, mains power is an obvious choice and the prime consideration then becomes the reliability of the supply.

In other cases where the site is in a remote area, it may be possible to extend a power line and in this case, the cost of the line will generally become the deciding factor. A high voltage power line, which runs close to a site, may be a tempting choice for primary power supply. However, the cost of installing a transformer and its associated protection devices could be very high. In fact, the operator of the line may not be willing to add the complexity of a drop point for the sake of a few kilowatts. In situations like this, it will usually be more practical to look to other sources of power.

5. .3   on essential essential and uninterruptible supplies

Power supplies derived from power lines and unduplicated generators are not regarded as being highly reliable, because these sources can be interrupted due to equipment problems, or the need for maintenance. Such supplies are generally only suitable for powering non-essential equipment, such as amenities and some percentage of the lighting load. Items, such as room heaters and electric jugs, should be regarded as non-essentials because of their high power consumption. In some installations, it will be practical to install a battery bank of sufficient capacity to operate the station for periods in excess of the expected mains power outage, but in general this will be limited to smaller installations.

An essential supply is generally a mains power line supported by a stationary diesel or other generating plant. If the power line fails, the site will be without power while the diesel engine starts up. This type of supply is quite adequate for communications equipment, which operates from battery supplies. The battery supply should be large enough to power the equipment for an extended period so that if the generator fails to start, there will be time to get a technician to the site to rectify the problem.

5. .4   Mains power station

In some situations the communications equipment, will require an ac power supply. Most computing systems are ac operated, as are many large radio transmitters. In such situations, it may be necessary to install an uninterruptible power supply (UPS), which generally consists of a battery bank, an inverter, and a battery charger.

UPS systems range from compact semi-portable units rated up to about 3 kVA through to large installations, which can go up to about 200 kVA and typically, these large systems use battery voltages of 400 volts. They can be single phase or three phase.

There are two basic types of UPS:

- Synchronous or on-line
- Off-line

The off-line types have a monitoring circuit that constantly checks the incoming mains for variations in voltage levels and interruptions. The battery charger maintains the battery in a float charge condition and the inverter is not running. When the mains go outside the specified limits, the inverter is started and a relay or solid-state switch connects the load to the inverter instead of the mains. This switching can be quite fast and modern systems can be on-line within half a cycle, which is acceptable in many cases. These types of UPS are relatively simple and inexpensive compared to on-line types.

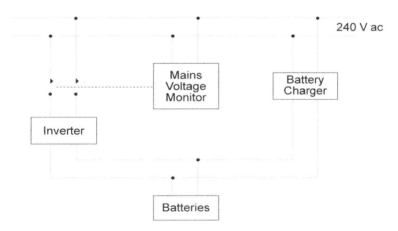

Figure 5.5
Off-line UPS

On-line inverters are generally similar to the off-line types but the inverter is running all the time and is phase locked to the incoming mains supply so that any variations in the supply will be taken up by the UPS. If the mains fail, it will be not evident to the load. There is a bypass circuit, which connects the mains directly to the load, and this operates when the UPS is isolated for service and for phase locking, during startup and in case of failure of the UPS. On-line systems are more complex and hence more expensive than off-line types.

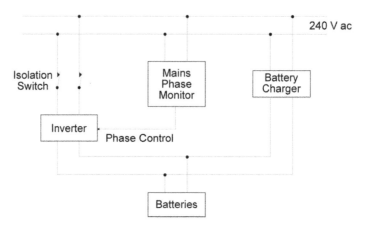

Figure 5.6
On-line UPS

Some of the points, which need to be considered in the specification of an UPS, are noted below.

- Carefully consider the time that the UPS will be required to operate under mains fail conditions. If a long operating period is called for the battery bank required may be very large. Generally, UPS systems should be used to carry the load while a standby diesel plant starts up and this will usually be about five minutes after the mains fail.
- In addition to the normal bypass switch within a UPS, it is a good idea to install an additional manually operated switch outside the UPS, so that it can be completely isolated for service and repairs.
- Sometimes the cause of mains failure will be a lightning strike, so be sure that the UPS is adequately protected against the very surges on the incoming power line, which may require it to operate.
- Ensure that the UPS can supply the inrush current of the equipment it supplies. Some equipment can draw two or three times the normal current when switched on.
- Ensure that the power factor of the load equipment is within limits, as the UPS inverter will not have the ability to handle poor power factor that the supply mains may have. Similarly, ensure that the UPS itself maintains an acceptable power factor.

5. .5   Solar power supplies

In recent years, there has been rapid development in the technology of manufacturing solar panels, which has resulted in a big increase in efficiency, and a corresponding reduction in the cost per watt. Typical panels are made of toughened glass and are highly resistant to rain, hail, abrasion by wind blown sand and impact. In short, solar power is now a very practical solution to the problem of providing power to small and isolated radio sites.

Some problems do remain, and the following aspects need to be considered in planning a solar installation.

- Solar panels must be installed so that they are unobstructed in their exposure to the sun. If the shadow of a support beam passes over the panel as the sun traverses it, the output will be severely reduced, as some cells are darkened.
- The panels may be vulnerable to vandalism or theft and may need to be well protected.
- The panels may have their output reduced by dust buildup in areas where little rain falls. Sometimes bird droppings can severely reduce the output and regular cleaning may be required.
- The installation needs to be carefully planned so that the panels are aimed with the correct inclination for the particular latitude. This information is always available from the manufacturers.
- The system designer must take into account that in some locations there may be many successive sunless days. Again, manufacturers will generally have good statistical data for the location.

Solar panels are generally manufactured in 12 volt modules and this allows modules to be stacked in-series to cater for 24 and 48 volt systems, and in-parallel to provide the current required, to maintain the station load and to recharge the batteries. The output voltage regulation is quite poor and typically a 12 volt module will measure over 21 volts on open circuit and this will drop to about 17 volts on full load. This allows sufficient cause for the use of a voltage regulator, which will maintain the output to a set point between 13.4 to 13.8 volts, which is the voltage range for lead acid batteries.

A typical modern solar panel will measure about 1200 mm by 550 mm, will have a normal output voltage of 12 volts, and will provide a current of about 4 amps.

5. . ind generators

In earlier years, wind generating equipment had a poor reputation for reliability. Wind generators required a good deal of regular maintenance and this was often expensive because, sometimes, riggers and cranes were needed.

They were very vulnerable to damage from wind and dust storms, and unless they were located correctly, the output was low and erratic.

Notwithstanding the problems, some installations have been operating satisfactorily for many years. Like solar power, wind power has recently benefited from a great deal of research and manufacturing development and it should be considered in suitable applications.

By far the most important consideration will be the wind and its reliability. In general, areas that are prone to cyclones and hurricanes should be looked at very carefully, unless these severe winds are so rare that the occasional loss of a generator would be acceptable. Obviously, areas that are noted for moderate to strong and regular winds will be ideal.

Having determined that a particular area is suitable, the system designer must next, carefully study the site and determine its suitability. Generally, flat areas are ideal, as the wind tends to blow evenly. Hilltops are sometimes unsuitable as the wind may be turbulent because of the up current off the side of the hill, which meets the direct stream at the hilltop. Similarly, small gullies should be avoided. At times, they can channel the wind to the generator's advantage, but if the wind is across the gully, severe turbulence will occur.

One of the big problems facing the wind generator manufacturer is the need to reduce the effective area of the turbine blades in high wind conditions. This was usually done with a feathering mechanism similar to that used in an aircraft propeller. This is an expensive solution and is mechanically complex. A recent approach to the problem uses a fixed blade propeller, which is faced into the wind by a tail plane. As the wind velocity increases beyond that needed to produce the rated output, the tail plane begins to tilt the propeller away from a 90 degree angle to the wind. This reduces the efficiency of the propeller and hence the wind force on it. If the wind reaches very high velocities, the tail assembly will tilt the propeller so that it eventually becomes horizontal, with both, the low efficiency needed to produce normal output and low wind drag to prevent damage, being affected. Table 5.1 shows some typical specifications for generators of the type described above.

RATED POWER			
At 6.5 m/s = 23.4 km/hr	300 W	500 W	2000 W
Power at 3 m/s	35 W	35 W	170 W
Max power at 8 m/s	540 W	800 W	2800 W
Rotor diameter	2.7 m	3.5 m	5.8 m
RPM	60–350	60–350	60–150
Weight	165 kg	205 kg	376 kg
Cyclone rated	Yes	Yes	Yes
Output volts	24/48	24/48	48
Relative costs	$6 K	$10 K	$20 K

Table 5.1
Some performance figures for wind generators

5. . Diesel generators

For many years, diesel generators provided the main power supply at isolated radio sites. This was especially true in the days of electron tube equipment, which required much more power than is common today. In many cases, the demand for power at an isolated site will be relatively small, and the big advances in solar and wind power generation mean that these sources can provide enough power for most conditions. Nature can be very difficult though, and sometimes, a prolonged period of rain with no wind may result in insufficient power to maintain the battery supply and in cases like this, a standby diesel plant will be essential.

Diesel plants can be very reliable but like any machinery, they require good maintenance and regular inspections to ensure that they will operate when needed. The design of the plant is equally important and some of the more important points are noted below.

- In older installations, it was common for a dc generator to directly charge the battery. These systems usually provided poor voltage regulation and often had no ac output. Modern designs use an ac output to power conventional mains regulated battery charging systems, and to provide power for all other station auxiliaries.

- The size of the plant is naturally determined by the load and it is important that a diesel engine is not operated on a light load, as this will cause the injector system to carbonize and the cylinder walls to become glazed, which will quickly cause problems. The plant should be specified so that the engine will run at about 70% capacity so that if the load were 3 kW the plant will be rated at 5 kW.
- In areas of high ambient temperatures, it is essential that the plant be properly cooled. Air intakes should not be located near air outlets and precautions should be taken to ensure that debris couldn't block these airways. When temperatures are very high, a thermostat is often placed in the control circuit so that the engine will run only in the cool of the evening but of course, the battery capacity must be able to carry the load for an extended period. This system has the added benefit of not charging the batteries when they are already hot during the day.
- In other areas, low temperatures will be the problem. In these situations, a standby plant may be very difficult to start because of the drag imposed by cold sump oil and it may be necessary to install sump heaters to keep the oil warm.
- In areas where sand and dust storms occur, it will be necessary to install highly efficient air cleaners, to prevent damage to the engine.
- If engines are in remote areas where maintenance may be irregular, it is wise to install a system, which will automatically maintain the correct sump oil level.
- All engines are subject to vibration and particularly in the case of single cylinder engines. This can quickly lead to problems such as fuel line fractures and equipment vibrating loose and hence falling off. Where engines are rigidly mounted to a large concrete base, it is likely that stress fractures will occur and if engines are mounted on shock mounts that are too flexible, vibration will be a problem. It has been found that thick hardwood beams provide an excellent mounting for engines allowing just a little movement to relieve the stress of a rigid connection to a solid base.
- Engines should be fitted with sensors to monitor temperature and oil pressure and should automatically shut down, if either condition becomes critical. These alarms, together with fuel level and fail to start alarms, should be included in a site supervisory system, to be monitored at the control center.

5.6.7.1 Standby plant

When a site is powered by mains, solar or wind generation, a single standby diesel plant would be used. The plant must be capable of running the essential load, which may include recharging the batteries as well as the normal station load. The diesel plant will normally start shortly after the mains power fails or in the case of a solar or wind powered site, when the battery voltage falls to a set level. It is important that a reasonably wide differential be set between the engine start and stop voltages, to avoid short running times, and in many cases a delay timer will operate to force the engine to run for a minimum period of about one hour.

5.6.7.2   Diesel wered installati ns

In some locations it will be impractical to install mains power and both solar and wind power will not be viable. In these situations, diesel generation may provide all the power required. As a rule, two generating sets will be installed, with one acting as a standby for the other but sometimes three are used where two alternate as the main and the third acts as the standby.

5. .   iltering dc supplies

One tends to think that because batteries are so large, they will prevent any interference from entering the supplied equipment via the dc supply – rather like a huge capacitor across the line. In fact, most batteries have a degree of internal impedance and at higher frequencies, they are not really very good capacitors at all. Generally, this will not be a problem as most modern equipment has an efficient input filter, but the system designer should be aware of the possibility of trouble when a dc supply is used to power other equipment, which may induce electrical noise into the battery supply.

Two interesting examples of actual problems are worth noting.

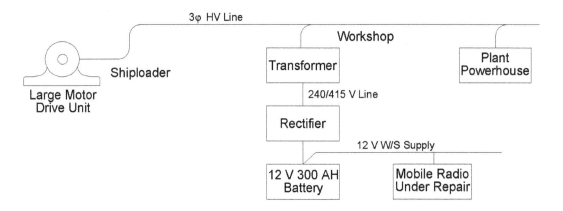

Figure 5.7
Case study 1

In Figure 5.7, a large motor, which operated intermittently, was fed from a plant powerhouse where the motor represented perhaps 10% of the powerhouse load. A general workshop, which also housed a communications workshop, was fed via a transformer and this supply was used to drive a rectifier and to charge a large battery, which was used to power up mobile radios undergoing repair. Very frequently, it was found that tantalum capacitors in the radios were being destroyed whilst the equipment was being serviced and the problem was eventually traced to high-level voltage spikes that originated from the motor, when it started. These spikes were not fully dampened by the rectifier or the battery. The problem was solved by installing a small ac motor/generator set to power the workshop equipment. The spikes were unable to conquer the mechanical coupling between the motor and the generator and this was a cheaper solution than the running of a new power feed for the workshop.

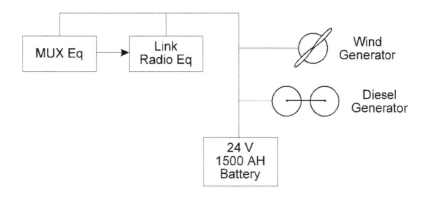

Figure 5.8
Case study 2

In Figure 5 8 a wind generator backed up by a diesel generating set was used to power a radio repeater station. It was often noted that a varying-frequency tone was heard in some of the telephone channels carried by the radio link. Eventually the problem was correlated with wind velocities and found to be caused by harmonic products of the ac ripple on the dc output of the wind generator. The battery system did not fully absorb the ripple and the multiplex equipment had poor input filtering.

The ac noise was modulated up into the baseband, fed to the radio link and thus entered some telephone channels. This problem was solved by installing an efficient filter network to decouple the multiplex equipment.

Modern equipment and good design practices will generally ensure that these types of problems do not occur.

5.7 Distribution (dc)

The system designer needs to give dc distribution systems the same attention, as he would ac distribution. The hazards of poor installation are similar and a badly designed system will be inflexible and unreliable.

The start point for a dc distribution system will almost invariably be one or more, battery banks. There may be a variety of different systems, typically 12, 24, 48 and 120 volts. Sometimes similar battery banks will be paralleled to provide increased capacity or reliability, and in cases like this, provision should be made to isolate a particular bank for maintenance.

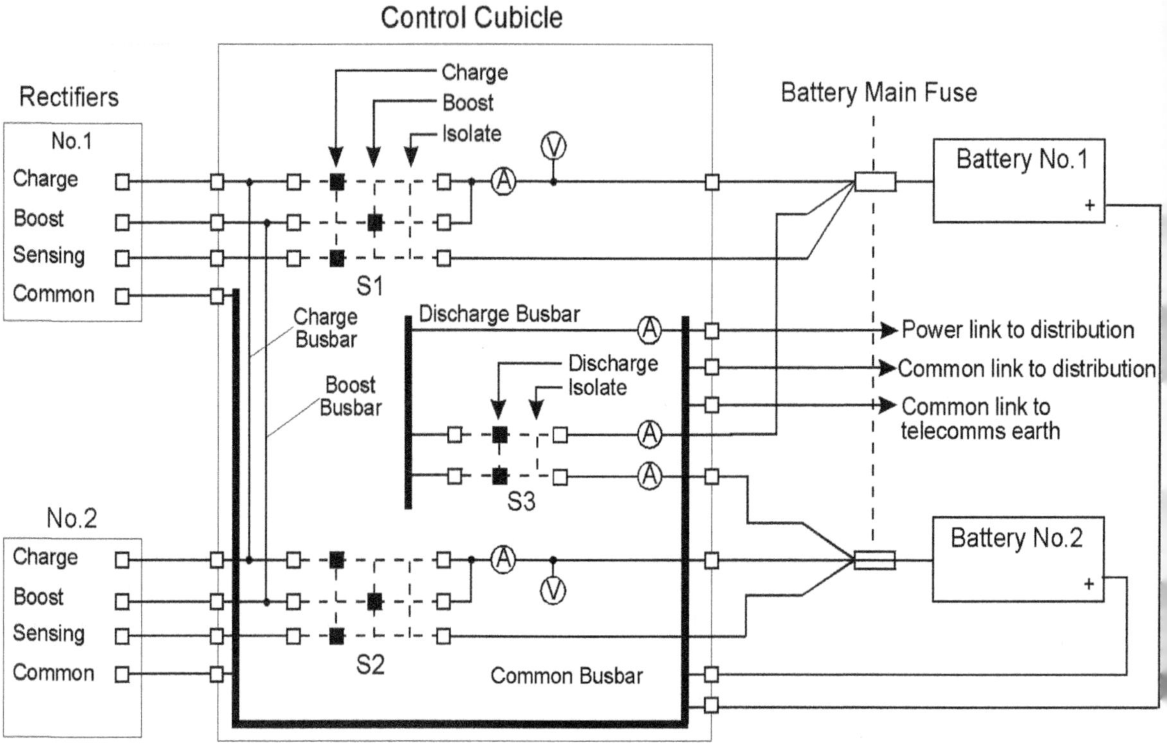

Note: Note how the rectifier voltage sensing circuits are connected directly to the batteries.
A charging system for the recombination cells would probably not have a boost charge.

Figure 5.9
A typical dc control schematic

A typical dc control system is illustrated in Figure 5.9 and some important features are noted below. The system is designed for a wet battery installation.

- Each battery bank has an associated rectifier and a three-way switch allows each rectifier to charge either battery or to be isolated for service.
- Each battery has a main fuse at the battery to protect the main battery feed cable and to allow complete isolation of a bank.
- The common (or earth) connection to batteries and rectifiers are brought to a battery common bar to which all equipment common lines are connected. This is to prevent any stray earth potential problems.
- The supply line from each battery is metered to allow monitoring.
- The supply lines are connected together at a main busbar, which is in turn connected to the main dc distribution system. This may use dc circuit breakers or fuses as required.
- The size of the battery cables is important, not only in so far as having an adequate current capacity, but also to minimize voltage drop. Some 12 volt equipment may draw 20 or 30 amps and at these levels, a very small resistance can easily allow a volt or so drop, across the distribution.
- Sometimes it is necessary to use relays to switch large dc currents. Conventional relays for dc service are large and expensive and can produce arcing, which may be a hazard in some situations. Mercury relays are not commonly preferred but have proven very reliable in use. A mercury relay

consists of a sealed cylinder that completely contains the hazardous mercury. Two contacts are set into the cylinder with one above the mercury level and one at the base. An iron sleeve floats on the mercury. When a current is applied to the actuating coil, the iron sleeve is drawn down into the mercury, displacing some of it so that the level rises to submerge the upper contact, thus completing the switched circuit. Mercury relays are capable of switching currents in the order of 100 amps and are virtually maintenance free.

5.8   Monitoring site alarms

In recent years, manufacturers have developed modern supervisory systems to simplify the operation and maintenance of their systems. This has been achieved through quality assurance programs aimed at improving reliability, and by providing effective remote diagnostics and controls, for their equipment. Today, almost all radio link and multiplex equipment can be connected to a service bus, which is carried as part of the radio traffic. At terminal stations and control centers, it is common to have a PC running software provided by the manufacturer, and by using this, a wide range of system supervision and management is available. Figure 5.10 shows a block diagram of a typical system.

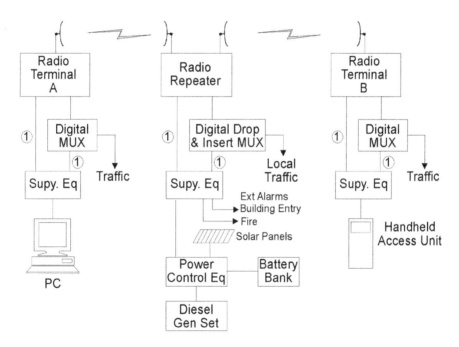

Figure 5.10
Supervision of a radio's system

At the terminal sites, a supervisory module is connected to monitor the radio and the multiplex equipment, and at the repeater site as well as the communications equipment, the module is also connected to external alarms, such as fire and building entry, as well as to the power supply.

At the A terminal there is a permanent connection to a PC while at the B terminal, a portable access unit has been plugged into the module to allow a visiting technician to check the system.

The functions and control facilities available to the maintenance staff will vary from vendor to vendor. The functions of a typical system are described below.

ccess to the system

From the PC, the user will enter an access code, which will carry information on their authority level. Some users may only be able to check system performance, while others will be permitted to change system-operating conditions. Once accepted, the user can enter a station identity code, and codes for the indication or control function required.

erformance

This level reports on the BER performance at the site. When there are problems on a long system, the BER can be checked at successive sites until the problem is located by degraded performance on a particular section. The BER will usually be reported in the following format:

LEVEL 1	Unavailable and total seconds	e.g. 11/2314
LEVEL 2	Severely errored and available seconds	e.g. 2/2303
LEVEL 3	Errored and available seconds	e.g. 13/2303
LEVEL 4	Degraded and available minutes	e.g. 4/38

Refer to Chapter 4 for further information on availability.

larm

This level reports on alarms present at the station. It is a simple matter to cable-in dry contact alarms, from station ancillaries, such as air conditioning, fire, mast lighting, power and security systems. The alarms can be indicated by a number or the software may be programmed to properly identify the alarm condition. The alarms can be reported over the system at all times, so that those that carry an urgent status, can alert staff at the terminal station.

GC Level

The AGC or automatic gain control is a measure of the strength of the received signal, and it is of great interest to the maintenance engineer, during periods when the link may be performing below specification. Typically the AGC will be calculated every five seconds and will be reported as a voltage level, e.g. −83 dBm. Sometimes the system software may be able to provide a continuous recording of the AGC, at a selected site, to assist in fault location.

Control

The control functions are usually password protected, as inadvertent use could shut a system down. Again, the control functions may be used to operate other equipment, and typically this will include starting and stopping the diesel engine, or perhaps turning on external lighting, so that staff who need to visit the site at night can see to unlock gates, etc.

There are many functions that can be carried out to test, or even reconfigure the radio and the multiplex equipment, and some of these are discussed below.

- Loopback tests

  If there are problems with a link and it is taken out of service, it is possible to address a repeater or the distant terminal, and to cause the receiver baseband output in the A => B direction to be connected back into the transmitter input, in the B =>A direction.

  By using this test, the system can be tested from the A terminal, with loopbacks applied at successive sites along the route, until the problem can be localized. It may also be possible to apply loopbacks within the multiplex equipment, to extend the range of tests to individual data or voice circuits.

- Reconfiguration

  At sites where there are drop and insert facilities, it is usually possible to change the configuration. Perhaps two telephone circuits were normally dropped at the repeater, in Figure 5.6, to connect to a PABX at Terminal A, to serve a local pumping station. Assume now that new pumps were to be installed and a contractor from a town, at terminal B, sets up a site office near the repeater. It would be possible for the supervisor, at the A terminal, to insert a data circuit and two additional telephone circuits, to go from the repeater site to the B terminal, and hence onto the contractors main office. Modern multiplex equipment often features electronic attenuators, which allow 'transmit' and 'receive' telephone levels to be adjusted remotely, to compensate for the physical telephone line conditions.

Some of the features of a modern supervisory system have been discussed, and it is obvious that a wide range of control and monitoring is available. This can be tailored to suit the particular system and will considerably reduce the need for routine maintenance visits, to remote sites.

5.9 Voice and data cabling – distribution systems

One of the greatest mistakes a system designer can make is to decide that a particular system configuration is final, and to install a cable network, which does not have any flexibility to allow future changes. The case where a group of telephone lines or data circuits is run from their source, directly to a multiplexer, is to be avoided at all costs, because if future changes are required, it will certainly require cabling changes and this in turn will probably lead to the interruption of other working circuits.

A general-purpose distribution system is shown in Figure 5.6 and this may serve as a model for other systems, be they large or small. The components of the system are discussed below.

Distribution frames   The   rone System

The early type of distribution frame, which used soldered tags to terminate the cables, has all but disappeared. It was inefficient in terms of space, and slow to terminate, and as the soldering skill of some workers now leaves much to be desired, it is probably unreliable.

A German company – Krone – developed an efficient, simple, and reliable terminating system, which with derivatives from other manufacturers, is now very widely used in cabling systems throughout the world. It uses an insulation displacement technique whereby the cable is inserted by a special tool between a pair of scissor-like contacts, which cut through the insulation and bend the cable into a **Z** shape, which grips it very firmly. When properly installed the incidence of termination problems in cables is almost unknown.

There are several types of termination blocks available and generally, each block will terminate ten incoming and ten outgoing pairs. The simplest is designed to join two cables without any monitoring or testing facilities. Another type allows for testing and interruption, so that an interrupt plug can be pushed into the block, to allow monitoring or testing in either direction. If a circuit is to be blocked out, a colored plastic plug can be pushed-in to break the circuit and the color can be used to indicate the reason for the interrupt. Other colored plastic plugs, which do not interrupt the circuit, can be pushed-in. These can serve to protect special circuits from inadvertent interruption, and again, the color can be used to indicate the circuit status.

The more common types of block are briefly described below.

- Disconnect blocks
  As described above
- Connecting blocks
  Allow connection of a monitoring cable, which do not disconnect the through circuit. These blocks also allow the installation of a high voltage protection module, which is widely used as lightning protection, on cables entering a building
- Commoning blocks
  Allow one incoming pair to be parallel connected to 19 outgoing pairs
- Instrumentation disconnection blocks
  Allow the connection of cables of larger conductor size than is common for communication cables
- Isolated cable blocks
  Allow the simple termination of a cable with no permanent further connection. Test plugs can be connected to the cable pairs
- Earth connection blocks
  Allow screen wires or earth common wires to be connected to a system earth point
- Screened pair blocks
  Allow the termination of a screen wire for each pair
- EIA-568 connection blocks
  Allow the connection of four pair circuits carrying voice and data signals in accordance with the EIA-568 Specification

All the above blocks may be installed, in combination on a series of mounting frames, ranging from a capacity of less than 100 pairs up to a maximum, imposed only by the available space.

The main distribution frame (MDF)

In general, there will be only one MDF at a plant or site, although sometimes there may be separate MDFs provided for telephone, data, and instrumentation systems. In most cases all the main cables for the site will terminate on the MDF and as the drawing shows, this includes cables to other buildings, to IDFs and to the outside world, via the public carrier system and a private radio system. All these cables will terminate on one side of the designated blocks, and all the pairs will be interconnected, as required, by jumper cables, which are run through marshalling guides.

It is of utmost importance that an accurate system of MDF records is setup, as the project is installed, and carefully maintained. Once the accuracy of the system is impaired, it becomes very difficult to re-establish it, and faults due to incorrect connections, will rapidly multiply. There are many systems for maintaining MDF records ranging from the original record book to modern software systems which are simple to use and administer and can provide multiple inspection of the records whilst limiting those with the authority to change them.

Intermediate distribution frames (IDF)

Figure 5.10, illustrates how IDFs are used as the next distribution stage after the MDF. They are generally smaller than the parent MDF and will be used to interconnect not only those circuits, which go back to the MDF, but also all internal distribution.

Final distribution frames (FDF)

In smaller buildings, FDFs are often not used and all distribution will be from the IDF. In office areas and other locations where a great deal of relocation of equipment takes place, the FDF can be very practical because, work can be done in a local area without having to re-run longer cables, which run back to a building IDF. As in the case for the MDF, it is most important to maintain good records of the frames.

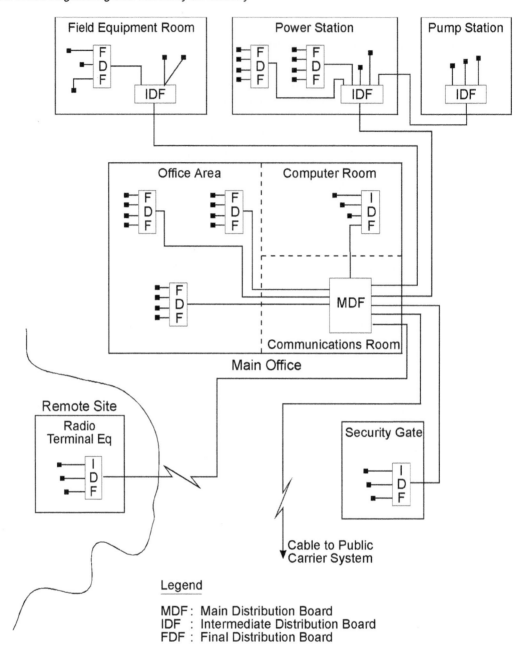

Figure 5.11
Example of a typical voice and data cabling distribution system

5.10   Equipment racks

In the past, most radio and multiplex equipment was manufactured to fit industry standard case sizes, and these were generally based on the British 19″ rack system. This is based on panel sizes, which are multiples of 1¾″ or 44.45 mm high and 19¼″ wide. Each increment is called a rack unit or 1 RU. Thus, a 2 RU panel is 88.9 mm high and a 6 RU panel is 266.7 mm high.

The 19" system worked well for many years and is still widely used, however, the dramatic reduction in size of electronic equipment has required many manufacturers to adopt alternative shapes and sizes for their equipment. This trend is reinforced, by the push to achieve high packing density for equipment, and because this type of equipment is often installed in city office buildings, where rates per square meter for floor space are extremely high, the annual cost of physically housing communications equipment becomes significant.

There is also a trend today, towards equipment, which is not permanently mounted into a rack, but instead sits on shelves in the rack; modems being a typical example. This allows units to be quickly removed for repair or to be installed elsewhere, perhaps where there is no rack-space. This is practical because, it is not necessary to have two versions of the equipment, but it emphasizes the trend away from the immaculate installations, which were something of a showplace for many operators, a few years ago. The AM broadcasting stations of the past were like technology exhibitions, with glass enclosures to house the high power RF stages, polished brass RF lines and mirror smooth linoleum floors. Installations like these are becoming rare, as equipment becomes more compact and floor space becomes more expensive.

The converse is seen in many computer equipment rooms, where equipment is placed on tables or cabinets in a completely unorganized way and a great spaghetti-like mass of gray cabling connects everything up. It is likely that such an installation will be prone to intermittent faults, as cables are frequently disturbed, and when it is a mess to begin with, there is little incentive for maintenance staff to look after it properly.

5.11 Interference in microwave and radio systems

Interference for microwave and radio systems is defined as any unwanted signal, which is injected into the traffic baseband, and with such a broad definition, the sources of interference will come from a wide range of origins. In most cases, the interfering signals will enter a radio system as some form of radiated signal.

Sometimes interference is caused by component failure. For example, a power supply filter capacitor may fail and high frequency noise from a switching regulator will enter the baseband signal. Problems like this are fault conditions and although they may be hard to isolate, they cannot be regarded in the same way as radiated interference.

Some of the more common types of interference are noted below.

5.11.1 igh voltage power lines

These can sometimes generate electrical noise, when insulators are dirty, and a discharge occurs. A broadband interfering signal is produced and it can easily cause problems, particularly with lower frequency systems. As the noise is likely to be intermittent, it may be difficult to locate.

5.11.2 Electromagnetic interference

Transformers and inductors radiate electromagnetic fields and if these are not properly screened within the device or the parent equipment, they will induce fields or currents into nearby equipment and this could cause interference. Generally, the standards authorities in a country will specify the permissible levels of electromagnetic radiation, which may exist outside the case of all electrical equipment, and these levels are usually low enough not to cause problems to other equipment.

5.11.3   Radar interference

A radar transmitter emits a high-powered pulsed signal into a high gain antenna, and because the antenna is rotating, the signal can be injected into communications links over a very wide area. Aircraft radar systems are amongst the most difficult to combat, and they are usually located on top of high mountains to give good coverage. Sometimes receiving antennas can be screened, but the system designer should be very careful to check the field strength and operating frequency of any radar system, near a proposed site for a communications link.

5.11.4   oreign system interference

Generally, the radio licensing authority in any country will be responsible for the allocation of working frequencies, but sometimes government authorities, large operators or the military may have groups of frequencies allocated to them. Again, the system designer should check to ensure that the frequency and field strength of any nearby transmitters are unlikely to cause any interference.

5.11.5   armonic and intermodulation interference

These types of interference have been discussed in Chapter 1. The problems are equally serious in radio link engineering and must be taken into account by the system designer.

5.12   Service channel

On most radio systems, except those of very low capacity, the manufacturer provides a telephone service for the maintenance engineer. Of course, in many cases, there will be adequate telephone facilities available on the site and such a facility may be of little use. Sometimes radio systems pass through a repeater station without being demodulated. There may be no traffic to be dropped from the baseband, or inserted into it, because there is no local town or installation. If a telephone service for the staff were needed and the baseband had to be demodulated and modulated again, it would involve a significant cost and in the case of analog links, would degrade the through traffic.

To overcome these problems, most radio terminals and repeaters have a small sub-baseband modulator and demodulator, which allows one or two voice channels to be dropped, and inserted, without any disturbance to the main baseband.

In the case of digital systems, the service channel will be a digital circuit and it will also allow for a drop and insert of the small supervisory system that reports station alarms, and allows for some commands to be directed to the station.

6

Integrating telemetry systems into existing radio systems

6.1   General

In order to evaluate the type of communications required for a telemetry application, it is important to consider telemetry systems in specific categories, i.e.:

1. **High integrity**
   High data throughput
   Continuous communications
2. **High integrity**
   Low data throughput
   Continuous communications
3. **High integrity**
   Low data throughput
   Intermittent communications
4. **Low integrity**
   High data throughput
   Continuous communications
5. **Low integrity**
   Low data throughput
   Continuous communications
6. **Low integrity**
   Low data throughput
   Intermittent communications

By placing a telemetry application into one of these categories, it is possible to determine the most appropriate and cost effective communications link solution.

Category 1 may require a quality microwave link, fiber optic cable or dedicated digital tie line service.

Category 2 may require a quality radio link, data quality copper cable or dedicated analog data tie line.

Category 4 may require connection into a LAN or an ISDN connection.

Category 5 may require connection into a radio link, on-site telephone cabling or analog tie line.

With categories 3 and 6, the application would involve a master site accessing the RTUs at regular intervals of greater than 10 minutes, or at irregular intervals, of anything from 10 minutes to several days.

For very remote RTUs, access via a normal switched telephone line would be sufficient. It is often more appropriate to implement this category of telemetry system over an existing radio system. The existing radio system is most likely to be used for mobile radios in vehicles. The telemetry communications equipment would be interfaced into the existing radio equipment and would have access to the voice channel, when it was not being used for mobile radio application.

The major advantage of such an implementation is the significant cost savings, in not having to design and install new master site radio equipment.

The following sections describe what is involved in implementing telemetry systems into existing radio systems.

6.2   Appropriate radio systems

The first step in evaluating an appropriate solution is to check if there is a radio system in use that is going to provide RF coverage to all proposed RTU sites. To do this, a thorough technical audit should be carried out on the existing master station.

The audit should, at a minimum, provide the following information:

- Age of equipment
- Condition of the equipment
- Output power of the transmitter
- Spurious output levels
- Type of antenna and cable
- Receiver 12 dB SINAD sensitivity
- Height of antenna on the mast
- Direction of antenna radiation pattern
- Latitude and longitude of site
- Average time for a voice call from beginning to end of conversation
- Maximum number of voice calls in one day (worst case)
- Spare power supply and battery capacity available
- Spare rack space available

Using this information, a full path profile should be carried out, from the master site to each of the RTUs. It was stated in this section that the minimum fade margin for a radio link should be approximately 30 dB above the 12 dB SINAD level. With the implementation of telemetry into an existing radio system, it may be difficult to achieve coverage of all RTUs to this level.

Considering the restraints of using existing equipment and that this is a low integrity system, it should be the design aim, in this case, not to drop the fade margin below 10 dB above the 12 dB SINAD level. If low receive signal levels require it, very high gain

YAGI antenna should be installed at the RTUs. If there are RTUs that cannot be accessed via the existing radio link, then consideration should be given to the use of RF repeaters (active or passive), possible access via landline or as a last resort, the use of a satellite station.

If a significant number of RTUs cannot be accessed by the existing site, then a cost analysis should be carried out to determine if it is more appropriate to set up a new master radio site with complete coverage.

Another important consideration when evaluating the suitability of existing infra-structure is the age and operational condition of the radio equipment. As a general rule, consideration should be given to replacing radio equipment every 10 years.

The installed lifetime of most base station radio equipment is 10–15 years. It is not recommended that new telemetry equipment be connected to old radio equipment. Some old radio equipment tends to exhibit a substantial degree of noise and distortion, reduced power levels, higher receiver sensitivities, highly non-linear frequency, amplitude and phase responses and spurious RF output emissions. This can reduce maximum telemetry data transmit speeds over radio to 300 or 600 baud. Some base station equipment of 10–15 years can appear to work perfectly at first sight, but it is always best to err on the side of caution.

Radio equipment that has not been maintained correctly can also exhibit some of the problems previously discussed. Depending on how the equipment has been mistreated will probably determine if it is redeemable or not.

6.3   Traffic loading

The next factor to determine is the availability of free channel time, on the existing voice channel, for telemetry data.

Communications that occur on a communications channel are referred to as traffic. The measure of traffic on a channel is referred to as the channel loading or channel occupancy. The maximum amount of traffic a channel can carry is referred to as the channel capacity.

With radio, one call of traffic should be considered a complete conversation. This may involve the two parties that are talking to each other keying up and down, two or three times each, before the conversation is complete.

The complete conversation may take one or two minutes, while the transmitters were only keyed up for three quarters of that time (quiet period as each party hands over to the other).

The aim then is to determine if there is any free capacity on the existing voice radio link for new telemetry data. To carry this out, the first step is to time a sample number of voice calls and calculate an average call time. Twenty to thirty calls should be timed and an average call time worked out. The result will greatly depend upon the base station's application. For example, a busy industrial site may make heavy use of the radio channel.

The next step is to count the number of calls that occur in the busiest hour of the busiest day for radio communications. Local users will know the best time to monitor the channel and carry this out.

Now the worst-case channel occupancy is calculated using this information gathered.

NOTE: It is always best to calculate traffic for a worst-case scenario so that when a telemetry system is interfaced over the top of the channel, the level of voice service does not deteriorate for the average user.

The following is used to calculate occupancy.

$$Channel\,occupancy\% = \frac{(No.\,of\,calls\,in\,1\,hour) \times (the\,average\,length\,of\,each\,call\,in\,minutes)}{60\,minutes}$$

For example, if the average call-length is calculated at one and one-half minutes and there are 22 calls made in the busiest hour, then:

$$Channel\,\,occupancy\,\, = \frac{22 \times 1.5}{60}$$
$$= \frac{33}{60}$$

$$= 55\,\%$$

The next step is to determine the worst-case channel time requirements of the telemetry system. The same procedure is used as described above. Obviously, if the requirement is greater than 45% channel occupancy then the system will fail (lock up). Therefore, the question is, what is a good, (acceptable) level of channel loading.

Calls onto a radio system are going to be statistically random in nature, i.e. they could occur at any time. Therefore, there will arise times when two parties try to access the radio system at one time.

In this case, there will be a collision and only one user will get access to the channel. The other situation is that while one party is using the channel, others may be waiting to access it. The term grade of service (GOS) is used to describe the quality of the communications channels, from a user's point of view, in being able to access the channel. GOS, as a figure, will give a statistical indication of the number of calls that will be unsuccessful in accessing the radio channel during the busiest hour.

It follows that:

$$Grade\,of\,service = \frac{No.\,of\,calls\,unsuccessful}{Total\,no.\,of\,calls\,attempted}$$

In simple terms, the grade of service can be calculated from the following formula.

$$GOS = \frac{T}{100 + T}$$

Where $T = channel\,occupancy\,in\,\%$

Using the previous example the grade of service will be:

$$GOS = \frac{55}{100 + 55}$$

$$= 0.35$$

$$\approx \frac{1}{3}$$

This means that one in every three calls will be unsuccessful. This is obviously a severely degraded system and a radio system with this GOS is well overdue for expansion. This was an extreme example.

It is sometimes difficult to determine what a satisfactory grade of service is, for a system. It depends on the application and on how long and how often users are willing to wait to access the system.

In general it is recommended that the grade of service not be greater than 0.2. That is, one in every five call attempts will be unable to get on the channel during the busiest hour of any day. Therefore, working backwards to find the maximum channel occupancy:

$$0.2 = \frac{T}{100 + T}$$

$$T = 25\%$$

It follows that total voice traffic, plus the total telemetry traffic, should not exceed 25% occupancy.

For important systems, a GOS of 0.1 or 0.05 is not uncommon.

The above overview of traffic analysis covered only the fundamental concepts (traffic analysis is a complete science in itself). Nevertheless, sufficient information has been provided to determine if a telemetry link will operate satisfactorily over an existing radio system, without causing call delay problems and upsetting existing users.

6.4   Implementing a system

With the more common types of mobile radio systems, a talk through repeater (TTR) type base station is installed. Here, all calls received from mobile radios are automatically retransmitted back out again and are heard by all other mobiles on that channel (assuming they are operating on the correct CTCSS frequency – refer to Section 1.21). Therefore, when a telemetry system is installed, every user will hear the telemetry audio coming from the RTUs and from the modem interfaced into the base station. This will be heard as a series of chirping tones that will be of great annoyance to voice mobile users.

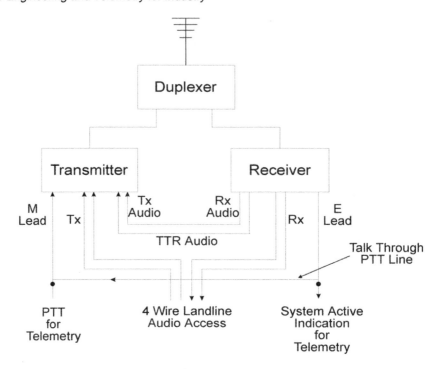

Figure 6.1
TTR base station with landline access

To overcome this problem there are two possible solutions. The first, and easiest to implement, is to give the mobile radio users one CTCSS frequency and the telemetry equipment a separate CTCSS frequency. When the telemetry radios are operating, the mobile receivers will not open and no telemetry signal will be heard.

The second method is to give every mobile radio and RTU radio a separate Selcal number (refer to Section 1.21).

This would enable the master telemetry unit to individually call each RTU and gather the required information back from them, without disturbing other mobile radios on the same system. This solution, although preferred to the CTCSS solution, may cost significantly more because the existing mobile radios may not have the Selcal option available, and special cards would have to be fitted to each mobile, to provide the facility.

The point at which the telemetry unit is interfaced to the radio is also an important consideration. Most modern base stations will have a minimum of one 4 wire interconnection point for landline access to the channel (this normally allows a nearby user sitting at a desk to get access to the system). If it is not being used, then the telemetry interface can be made at this point.

E and *M* signaling leads are provided for controlling the base station. When the *E* lead is active, the base station is in use, which indicates to the telemetry unit that it cannot get access. When the telemetry unit wishes to transmit, it activates the *M* lead which keys up the transmitter (assuming the *E* lead is not active). The *M* lead is often referred to as the press to talk (PTT) lead.

If a different CTCSS tone is required for the telemetry system, then a separate connection will be required into the base station to activate the CTCSS tone for the telemetry radio group, and in some cases a new CTCSS card may be required for the base station.

If Selcal tones are to be used, these will generally have to be generated by the telemetry unit. This will require extra hardware and software in the telemetry unit. Selcal circuit cards can be purchased and programmed to produce the Selcal tones required. Some modern base stations have RS-232/422/485 digital interfaces through which Selcal and CTCSS tones can be controlled and automatically adjusted.

The landline interfaces, in base station radios, are normally 600 ohms impedance, with option for two wire or four wire connections. The required input level will vary considerably from different manufacturers.

On older radio equipment, if there are no landline interfaces, then connection can be made through the microphone input (care must be taken to match the impedances and transmit and receive levels).

Finally, the RTU radios must also be carefully selected so that the correct Selcal and CTCSS facilities can be included.

6.5 Trunking radio

A trunked radio system is one in which two or more radio channels are time shared by a larger number of user groups on the system.

The trunked radio system uses a control channel, for transmitting digital control information, through which a user group can gain access to one of a pool of available channels. The system is very similar in concept to a switched telephone network. Rather than dedicating a single channel to a user group, the resource of channels are distributed for use among a greater number of users.

For example, there may be four separate mobile radio user groups that generally do not need to communicate with each other and two separate unrelated telemetry systems, all of which are to operate from a single site three-channel trunked radio system. The digital control channel is used to allocate one of the three channels to a user group when they require communicating.

Each user group will hear only the calls intended for them because of automatic selective calling built into the system.

Trunking radio systems are very sophisticated and offer many additional facilities and options. Interfacing telemetry systems into existing trunked radio systems can be an excellent solution, provided there is sufficient capacity on the channels. The same traffic calculations would have to be carried out as were discussed in Section 6.3, but in this case for multiple channels.

7

Miscellaneous telemetry systems

7.1   Introduction

Three different telemetry systems, which differ from standard telemetry systems, are discussed here. The communication media, range from acoustics, infrared to inductive coupling and modulated backscatter. Although they are possibly out of the mainstream of telemetry systems, they are nevertheless important as they provide solutions, where traditional approaches would not be adequate.

These techniques will be discussed with reference to real situations to emphasize the practical nature of the telemetry application.

The telemetry applications that will be discussed are:

- Ocean data telemetry (using cable/acoustic methods/inductive coupling as the transfer mechanism)
- Physiological telemetry (with radio and infrared media)
- Vehicle ID information (using modulated UHF backscatter)

7.2   Ocean data telemetry application

The main advantage of telemetry systems, as applied to ocean data, is the ability to access the data in real time, for operational and forecasting purposes. There are a few low baud rate satellite telemetry systems, currently being used with low power radio transmitters, using omni directional antennas based on ocean buoys. These are most appropriate for transferring data from surface-to-shore and provide satisfactory results. Two well-known satellite systems here are the Argos Data Collection System and the Geostationary Operational Environmental Satellite Data Collection System (GOES). There are not many satisfactory methods available for transferring data from the depths of the ocean to the surface, mainly due to the harsh environmental conditions and the fact that seawater is effectively the intervening medium. The typical parameters transferred by the transducers to the surface are current (via a current meter) and temperature.

The three methods of transferring the data from the ocean depths to the surface are discussed in this section. These are based on the work done at the Woods Hole Oceanographic Institution in the USA.

7.2.1 Electromechanical cable

The first is the use of electromechanical cables as both the strength member and the communications medium. Figure 7.1 gives a diagram of the electromechanical cable.

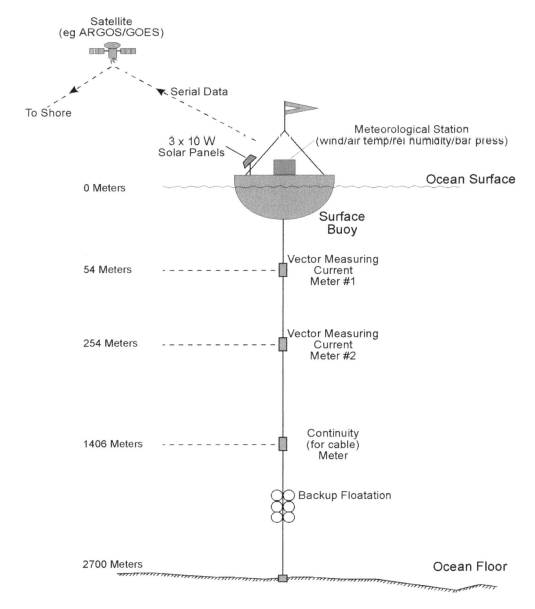

Figure 7.1
Electromechanical cable

The specially strengthened electromechanical cable formed a basis for this design, with the inner section containing three stranded copper conductors, laid in the grooves formed by the three main wire strands. Besides the corrosive impact of seawater, provision has to be made for fish biting the cable. The other possibility, of using parallel cables secured to each other (one for strength and support and the other for the electrical conductors), was rejected because of difficulty of deployment into position and the cumbersome overall cable structure.

The full system is indicated in Figure 7.1. A controller is based on the buoy and requests data from each of the instruments at various intervals.

FSK communications were used for communication to each of the current meters.

Typical problems that were experienced:

- Wiring errors in cable terminations on the cable itself
- Corrosion due to different metals making contact on the cable
- Software problems and failure of the FSK modem on one of the underwater instruments

7.2.2   Acoustic modem

The second approach is using acoustics, for transferring data between the subsurface instruments and the surface buoy. The information is transmitted and received, using acoustic transducers. There are two types of modems that were used:

A low power unit (based on the Motorola 68 HC-11 microprocessor) operating primarily as the transmitter at 1200 bps FSK. Receive speed is 100 bps FSK.

Table 7.1 lists the specifications of the two types of modems:

PARAMETER	LOW POWER MODEM	HIGH BAUD RATE MODEM
Frequency range	15 to 20 kHz (data) 12.7 kHz (command)	15 to 35 kHz (data) 13 to 14 kHz (command)
Modulation technique	16-tone MFSK (data) FSK (command)	256-tone MFSK (data) FSK (command)
Data rate	1200 bps (data) 100 bps (command)	4800 bps (data) 100 bps (command)
Range	5000 meters	5000 meters
Acoustic power	186 dB	186 dB
Electrical power	10 watt (transmit) 0.3 watt (receive) 0.01 watt (standby)	10 watt (transmit) 3 watt (receive) 0.01 watt (standby)

Table 7.1
Specifications of two types of acoustic modems

A polled philosophy is used where the high-speed modem is used to communicate to multiple low power modems. An alternative philosophy would be for several of the remote modems to communicate back on a predetermined time schedule. Each modem is assigned a unique address.

The low power modems remain in standby mode (at a low power consumption of 0.01 watt) until a command is transmitted from the surface modem to the appropriate address, which then activates the modem. The standard FSK modulation technique is modified to transmit on a number of different frequencies, as the bandwidth is far greater than a normal telephone system, for example. MFSK (or multiple frequency shift keying) allows many data bits to be transmitted at the same time thus allowing for a far greater data rate.

The technique of operation of MFSK is shown in Figure 7.2. digital signal processing (DSP) technology has to be used to decode the range of frequencies.

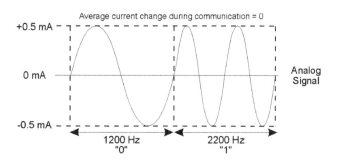

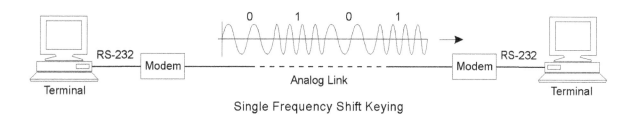

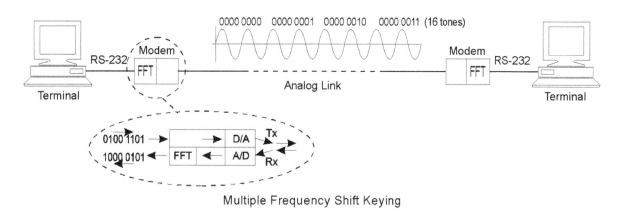

Figure 7.2
Use of multiple frequency shift keying

When uncoded transmissions (without any error correction) are used, BER of the order of 10^{-3} to 10^{-4} are normal. In using error correction this improves to 10^{-4} to 10^{-5} (almost a factor of 10).

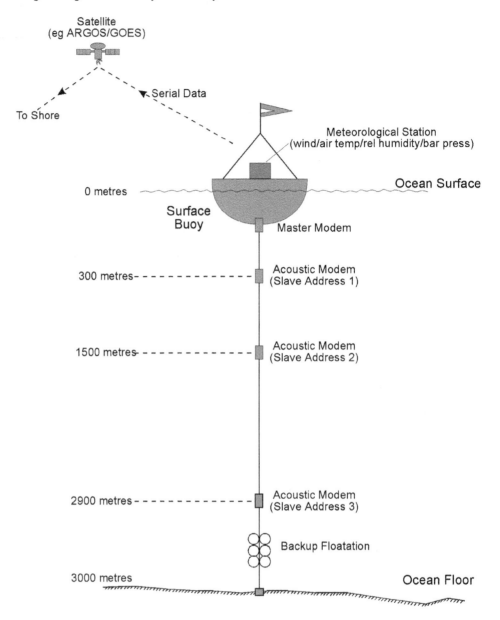

Figure 7.3
Telemetry system using acoustic modems

Interestingly enough, shallow water is the most difficult medium due to multipath reflections on the surface and the ocean bottom. The extraneous sources of noise from the surface and reflections on the surface can be minimized by the use of a sound baffle at the receiving modem.

7.2.3 Inductive modem

The third approach is to use an inductive modem approach to transmit data from the remote instruments. This offers a low cost approach to acoustic or cable systems although the energy efficiency is low. Distances of communication are of the order of a few thousand meters. Essentially a frequency shifted key signal is induced into the cable, which transmits it to the receiving modem.

Figure 7.4 gives an indication of how the inductive modem is constructed.

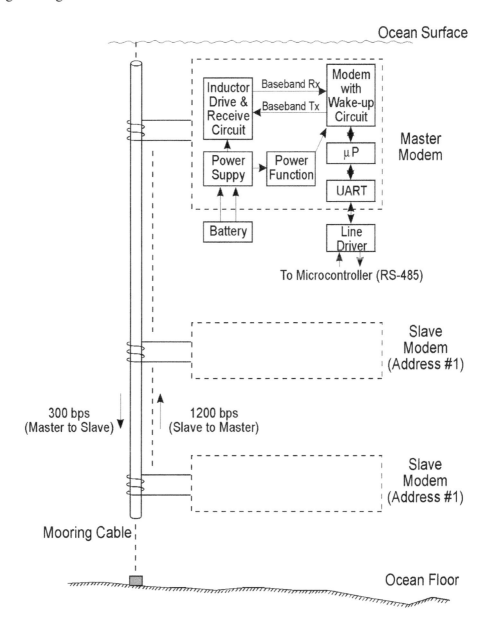

Figure 7.4
Operation of the inductive modem

The single chip modems used for the telecommunication environment are appropriate for inductive applications as the insulated steel mooring wire immersed in salt water has an attenuation versus frequency response quite similar to twisted pair telephone cable. The total system is setup with each remote modem being assigned a unique address. The link operates at 1200 bps from the remote modem to the master unit transferring the data and 300 bps from the master to the remote.

The transducer of the inductive modem is in the form of a toroid, which clamps around the cable at the transducer point.

Typical specifications of the inductive modem are:

PARAMETER	DESCRIPTION
Frequency	1200/2200 Hz uplink 2025/2225 downlink
Modulation	FSK
Baud rate	1200 Uplink 300 Downlink
Power requirement	350 mW active 5 mW quiescent mode
Estimated range	10000 meters

Table 7.2
Inductive modem specifications

7.3   Physiological telemetry application

There is often a need to measure physiological data remotely using some form of telemetry system. This telemetry system records data from non-human primates (baboons) and is used to study their behavioral patterns. This section discusses the use of a telemetry system based on a radio system for transmission of arterial blood flows, arterial blood pressure and heart rate back to a central station. An infrared link is then used to transmit back command information (to the animals) to control the animal's radio transmitter and to control the drug sources for each animal. Figure 7.5 below gives a diagram of the overall system.

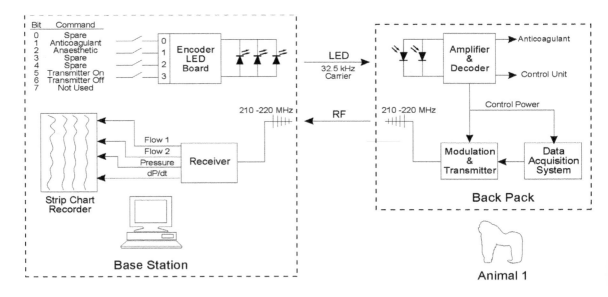

Figure 7.5
Overall physiological telemetry system

The required specifications are described in the Table 7.3.

DESCRIPTION	SPECIFICATIONS
Number of analog channels	4
Bandwidth per channel	0 to 50 Hz
Carrier frequency of RF system	210 to 220 MHz
Range of RF system	20 meters
Carrier frequency of infrared system	35.5 kHz
Range of infrared system	10 meters
Humidity	0 to 90%
Ambient temperature	0 to 40°C

Table 7.3
Specifications for physiological system

Each baboon wears a backpack containing a telemetry transmitter, infrared receiver and the associated transducers. The baboons are contained in a large cage and their behavioral patterns observed. The data acquisition telemetry (radio frequency) and the command systems (infrared) will be discussed separately in the following.

The radio frequency system handles the four transducers for measuring blood flow and blood pressure. The telemetry link uses a fixed frequency FM transmitter and receiver direct modulated by a serial data stream. As discussed earlier four analog channels (plus an additional housekeeping channel) are multiplexed on the serial data stream.

The transmitter module operates within the frequency range of 210 MHz to 220 MHz. The transmitter antenna is a spiral half-wave loop printed in gold plated copper. A copper counterpoise, which covers a second board, can have the separating distance between the two antenna boards varied to tune the transmitting frequency (between 210 MHz and 220 MHz). A single tuning capacitor (7 to 40 pF) connected from the center of the antenna to the counterpoise is adjusted to match the antenna to a 50 W transmission line. The standing wave ratio should be less than 1.2 to 1 over the frequency range of 2 MHz.

The receiving antenna array consists of four loop antennas. The loop is built on an aluminum counterpoise and is fitted with a 10 dB in line amplifier.

The command module is based on infrared and is used to transmit instructions to the baboon back pack for control of a syringe for administration of drugs or power on and off to the system's transmitters. Eight digital signals are controlled using a subcarrier frequency of 35.5 kHz. The subcarrier is chosen to minimize the power consumption and is encoded with a serial code of nine bits.

The system worked well apart from a problem with radio frequency drop-outs. This results from nulls produced in the standing wave pattern, produced by the transmitter and the cage. These nulls change depending on where the animals are situated in the cage. As the cage walls are not perfect reflectors the minima are not true nulls. Hence it is possible to reduce the number of communication drop-outs by increasing the power of the transmitter, the gain of the receiver or (more difficult) using the concept of diversity reception, with multiple frequencies used on the data acquisition telemetry link.

7.4   Tag identification system using modulated UHF backscatter

An interesting telemetry application, with passive transponders using modulated backscatter, is discussed below. The standard approach for tag identification systems is to use optical labels with some form of barcoding to encode the data. However, they are not very effective when there is some obstruction in the way or the barcode label and the reader cannot be accurately aligned (especially when the barcode and the associated reader are moving with respect to each other). The alternative is to use some form of radio frequency tagging system. This would be especially useful for vehicle identification systems, where there could be fairly rapid movement between the vehicle tag and the device reading the tag.

There are various approaches open to using some form of radio system to transmit the unique data contained on the tag. This could be a radio transmitter embedded in the tag, actively transmitting to a radio receiver. An example of this active type of tag is the *downed airplane beacon*. They are not used much in general industrial applications, due to the high cost and additional complexity of maintenance. The general practical electronic tag is the one, which is energized by a carrier signal sent from the reader and then returns a signal derived from the energizing signal.

There are two types of radio frequency tag systems:

- Inductive or near field (60 kHz to 300 kHz) use loop antennas with a range of half that of the longest dimension of the reading loop
- High frequency or far field (100 MHz to 5 GHz, say) use antennas such as dipoles, dipole arrays or horns with ranges much larger than the antenna dimensions

The system that will be examined is one that is used effectively for railway wagon identification purposes. Frequencies that are chosen in the USA for this application are 915 MHz and 2450 MHz (but applicable to most locations around the world).

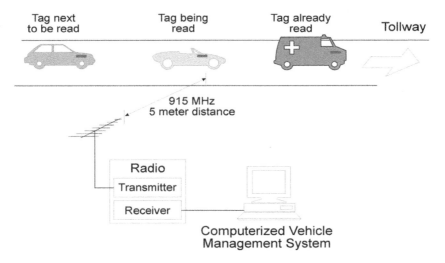

Figure 7.6
Practical application of tagging system

The reader transmits a continuous carrier wave to the tag. The tag then reflects (or backscatters) some of this signal back to the reader, in an amplitude modulated form, corresponding to the tag data. The reader then decodes the returned signal, as in a homodyne Doppler radar receiver (uses a single frequency), and extracts the tag message.

The tag consists of three elements:

- An antenna resonant at the carrier frequency
- A controllable impedance to modulate the backscatter
- A circuit to generate the message signal corresponding to the tag data

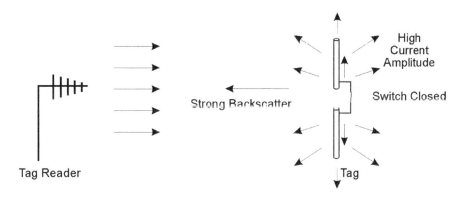

Resonant Antenna with Strong Backscatter

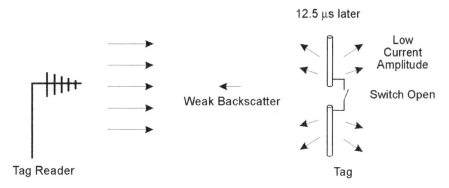

Resonant Antenna with Weak Backscatter

Figure 7.7
Principle of operation of backscattering

The operation of the system, to generate an amplitude modulated wave, is fairly straightforward. When the two halves, of the halfwave dipole antenna connected to the system, are resonant, they result in a reradiated signal scattered in all directions, at the maximum signal intensity. If the dipole terminals are opened, the separate quarter wave sections are no longer resonant and a much smaller RF current flows, resulting in a lower intensity reradiated signal. Hence, the circuit that needs to amplitude modulate the intensity of the signal, only needs to encode the appropriate tag information on the RF carrier. It does not need to generate the antenna power or modify the carrier frequency. Hence, the power consumption is minimal. There is, however, a choice between the use of battery powered tag antenna systems or carrier powered. The ranges of the two systems vary. For a specification of 2 watts of RF radiated power at 915 MHz and a 10 dB antenna gain, the working range is of the order of 9 m, for a passive powered beam, and 23 m for battery powered active tags.

On balance, battery powered tags are probably better than beam powered antenna systems; as the lifetime of a battery is in the order of five to eight years, for lithium cells. This figure can be doubled with better design and more appropriate lithium cells. In an application where the distance may be a critical parameter it is preferable to use battery powered systems, as the increased radiated power required, may need to be increased by four times to obtain a two fold increase in distance (in accordance with the inverse square law of quadrupling radiated power energy for doubling distance).

The protocol structure for the coded tag message is based on a 128 bit message with the two bits designated by 20 kHz and 40 kHz square waves as indicated in Figure 7.8.

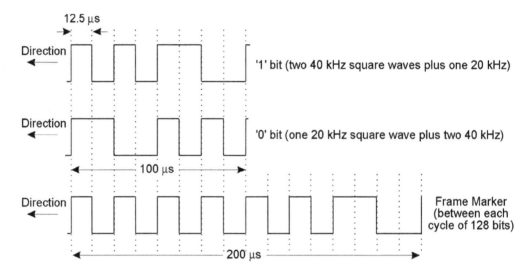

Figure 7.8
Coded tag signal

The reading speed is dependent on the time it takes to receive one complete code frame, while the tag is within range. Typical theoretical speeds are about 185 km/h whilst the French TGV train has its tags read at up to 280 km/h.

One problem that requires attention, in the receiver circuitry, is that the phase of the return signal varies with respect to the reference signal, as the distance between the tag and the detector systems change. The receiver detector output is proportional to the cosine of the relative phase between the two RF signals; the relative phase changes being dependant on the number of wavelengths for the round trip distance. When the phase difference is either 90 or 270 degrees, the detector output is zero. A phase difference of 90 or 270 degrees presents problems in that the detector output will have a zero output, thus giving no information about the status of the tag. A solution to resolve this problem is given in Figure 7.9 below:

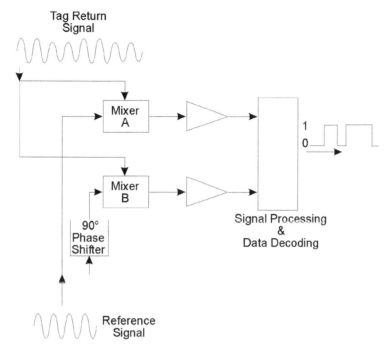

Figure 7.9
Phase shift correction circuit

This detector circuit has two double balanced mixers, which operate 90° apart, in reference to the return signal phase. This approach guarantees that at least one good signal is always produced, regardless of the return phase of the RF signal.

The applications for the backscatter tagging approach are numerous, ranging from tagging of individual railway cars to vehicles traveling over a toll-way. They allow the automatic reading of vehicle information without operator intervention.

For this tagging system to gain universal use there must be a standard developed for the RF frequency, message length, message content, signal strengths and operating speeds.

8

Practical system examples

8.1   A dockside communications system for LNG tankers

Liquefied Natural Gas is transported in specially designed product tankers, which are easily recognized by the characteristic domed storage tanks that project above the deck of the ship. The gas is in a liquid state and at a temperature approximately −140°C. It is an extremely dangerous product to transport and stringent safety procedures must be observed. When an LNG tanker is loaded, large articulated pipes are connected between the tanker and the loading jetty, and the flow of the LNG is controlled by the onshore process control system, which is linked to the tanker control system.

Should any anomaly occur it will be detected by the control systems and should an emergency arise, an emergency shutdown (ESD) system will abort the loading process.

- **The problem**
  To provide a high reliability link between the LNG tankers and the onshore process control system. The conventional solution would be to use an electrical cable that could be plugged into the tanker control system. In the past, unloading ports had used a fiber cable connection. For several reasons the use of a cable system had become unacceptable.

- **The solution**
  A 2 Mb radio terminal was installed near the loading jetty, as the *A* terminal of a single hop radio link. A similar radio was installed on each of the seven LNG tankers, as the *B* terminal, and an eighth identical *B* terminal was installed at another onshore location to act as a link monitoring station, and as a working spare equipment, which was designed to allow quick replacement of a faulty tanker terminal.

  The transmitter on each tanker is normally turned off so that it cannot cause interference in other ports and so that, if two tankers were in port at the same time, the waiting tanker could not cause interference to the one at the loading jetty. When a tanker approaches the port, the onshore loading controller presses a tanker select button, which sends a coded signal to the tanker receiver, and if the code matches, the tanker transmitter is turned on.

The 2 Mb link is then available for traffic and several telephone lines on the tanker are available for the ship's business and for the crew to call friends and relatives. This is a useful feature of the system because it may take an hour or so, for the tanker to be securely berthed and a conventional cable link established. The process control and emergency shutdown links are also available, and may be tested so that loading may proceed as soon as the loading arms are connected.

The radio connection to the tankers has another important advantage. If a serious problem were to develop on the loading jetty, or if a cyclone developed, it would be necessary for the tankers to leave the jetty immediately. In this case, there is no cable connection to worry about and important telephone and data links are maintained during departure.

When a tanker leaves the port, the control operator deselects the tanker and the system defaults to the onshore *B* terminal, which upon receiving the correct control code, begins to transmit, and sets up some data loopbacks, which monitor the system all the time.

The radio equipment at the *A* terminal is fully duplicated and operates in the cold standby mode. For reasons of cost, the tanker radio equipment is not duplicated. Should a major fault develop with a tanker *B* terminal before or during ship-loading, the spare onshore *B* terminal can be quickly installed on the tanker, to allow loading to proceed.

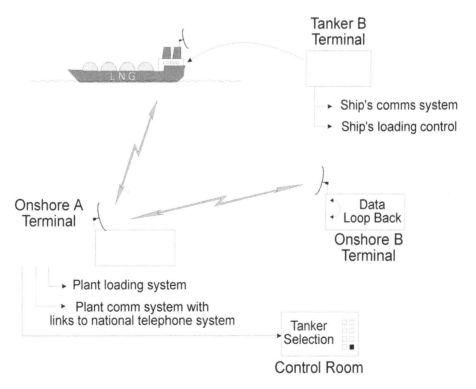

Figure 8.1
A dockside communications system for LNG tankers

8.2   Remote oceanographic sensor system

The problem

To be able to obtain information from a remote oceanographic sensor system, e.g. a tethered buoy in the Indian Ocean, on the edge of the Australian Continental Shelf, 150–200 km from land. The information is to be in digital form. The size of the buoy is 2–3 m in diameter.

The information to be gathered includes:

- Ocean currents
- Sea temperature
- Humidity
- Wind speed and direction
- Wave direction
- Wave height
- Wave period
- Tide information

These are but a few possible pieces of information, an oceanographer, or a weather forecaster require in performing their duties.

The solution

There are several solutions to the above problem. Appropriate ones will be discussed briefly below.

a. Stored data

The information being gathered by the sensors on the buoy could be stored in an onboard data collection system, i.e. magnetic tape. The stored information could then be collected by periodically changing out the tape in the buoy, 3–4 times per year, at which times, maintenance could also be carried out on the buoy, making it a more cost-effective exercise. This method, of data collection and storage, is used successfully where the data is for long-term study, and not required during or just after a particular event, i.e. cyclone or winter storm. If the data is required regularly, a more cost-effective method (other than frequently traveling to the buoy) needs to be employed.

b. Transmitted data

A more successful solution would be to transmit the information from the buoy to a shore station where it could be received, stored and processed in real time, giving the end user instant access to the information, even during an event, as mentioned above.

There are several possible ways of transmitting and/or communicating information between the buoy and the on-shore station, by using either HF, VHF–UHF, microwave or satellite transmission.

- **VHF-UHF microwave transmission**
  The distance stated in the problem (150–200 km from shore) virtually eliminates the use of VHF-UHF and microwave because over this distance, line of sight is required to give reliable communications, and unless a very tall tower or a high mountain is available onshore, this cannot be achieved. Tall

towers are very expensive, and as tall mountains are scarce around the West Australian coastline, the two remaining methods are HF and satellite transmission.

- **HF transmission**

  HF transmission is an effective method of communication but requires efficient antennas to give reliable communications. At the distance given in our problem, frequencies in the order of 2–3 MHz would be required to give reliable communications. At these frequencies, the antennas would need to be quite large (8–10 m long). As the size of the buoy is limited, it is virtually impossible to construct an efficient antenna of this size. However, there are oceanographic buoys that do use HF but usually operate in the 10–14 MHz range and the buoys are very large (5–6 m in diameter), allowing for efficient antennas to be constructed on board. To overcome the inefficiency of the antenna, the transmitter power could be raised, but more power means more drain on the batteries, which would necessitate having larger batteries. To overcome the battery problem, the data could be compressed or reduced, which would probably be undesirable, as the end user usually prefers as much data as possible.

  HF transmission does have other problems, such as the vulnerability to interference (corruption of the data) from such sources as other HF users, man-made noise (electric motors, welders, switch gear, etc) and electrical storms. Because of these problems, special encoding of the data is required so as to pass the information, without data being lost (corrupted). The speed at which data can be passed over HF is limited by the bandwidth available, and is normally 300–600 baud, although with modern and more expensive modems, higher speeds (2400 baud) are possible. The faster the data is passed the more vulnerable it becomes to corruption. Hence, the reason for the higher cost for the modern high-speed modems.

  The slower the data rate the longer the transmitter has to remain on, and therefore, the greater the power drain on the buoy's battery system. The more power that is required, the bigger the batteries have to be, as previously stated above, and therefore, the greater the weight, which leads to buoy instability and other problems.

- **Satellite transmission**

  To communicate to and from the buoy a modem will be required to encode and decode the information to and from the onboard sensors. The communications path from the buoy to the shore station could be via several satellite systems.

Orbiting satellites

The information from the buoy could be passed to any one of the many systems in use, such as the *Argos* system, which uses low orbiting satellites. There is difficulty with this system in that the satellites are not always present above the buoy. This leads to periods where no information can be either sent or received from the buoy. Due to this situation, when the buoy cannot pass information, it must store it and then send it when a satellite window is available.

The information that is sent to the satellite is then returned to an earth station, where it is passed on to a bureau, where the information can be collected later, via leased telephone or data lines. This system adds extra expense because of the need to store information on the buoy, and the added sophistication of the satellite communications transceiver on the buoy.

Geosynchronous satellite

The information from the buoy is passed to a geosynchronous satellite, which in turn passes the information to a ground station for relaying on to the end user. By using a geosynchronous satellite, the buoy is able to pass and receive information at all times, because the satellite can be *seen* by the buoy at all times. Such a system can be provided by the Inmarsat Service.

This system divides the world into four zones:

- **West Atlantic Ocean**
- **East Atlantic Ocean**
- **Indian Ocean**
- **Pacific Ocean**

In Western Australia, which borders onto the Indian Ocean, both satellites, i.e. Indian and Pacific, can be seen, giving a great advantage to be able to pass traffic on either, when congestion may occur. The Inmarsat system can and does operate through the OTC Telstra Ground Station at Gnangara, about 30 km north of Perth, which gives the system direct access to the public data switched network and telephone network.

As well as the ground station in Australia, there are many others in the South-East Asian rim, i.e. Singapore, Japan, China, India. There are other stations throughout the world, which can be accessed.

Brief description of geosynchronous system

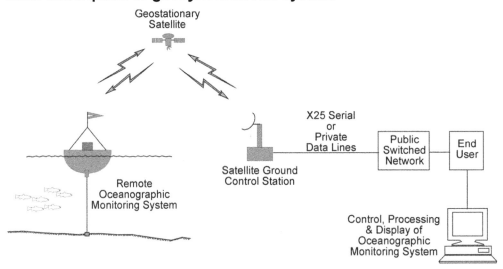

Figure 8.2
Remote oceanographic sensor system

A remote oceanographic monitoring system, which is a tethered buoy in the Indian Ocean, gathers information from its sensors and stores it in an onboard computer. At a predetermined time, the satellite transceiver is woken up by the computer. The transceiver then establishes a communications link from the buoy, via the satellite, to the Gnangara ground station. Once the link has been established, the buoy then passes its gathered data to the ground station. When the buoy has completed its transmission the computer puts the transceiver back to sleep, thereby saving power.

At the ground station, the received information is stored in the station's computer. From the preamble codes that were transmitted from the buoy, the computer then, either, passes the information via the public switched network, or via private lines, to the end user. The time elapsed from the arrival of the information to the delivery to the end user is approximately 2–3 minutes.

NOTE:

The system is a *store and forward* data system: that is the information is passed from the buoy or from the end user, to the ground station, for onward forwarding to either party. At no time is there direct contact between either party. The system is therefore not a real time system.

When the information reaches the end user's site, it is stored, processed, and then displayed, as required, on a computer network.

If at any time the end user desires to reformat the buoy's sensor collection process, this can be done by sending a command signal to the buoy via the main station. This allows the end user flexibility in gathering the information needed, and the time it is to be gathered and returned and retransmitted to them.

Because of the sophisticated coding and decoding of the signals passed over the Inmarsat network, 100% reliability, and confidentiality of the data, is guaranteed. The data speed is 600 baud and is compressed to save time of on-air transmissions.

It should also be noted that the system is available at all times and through all weather conditions, which is a desirable feature when gathering data in stormy weather and high seas.

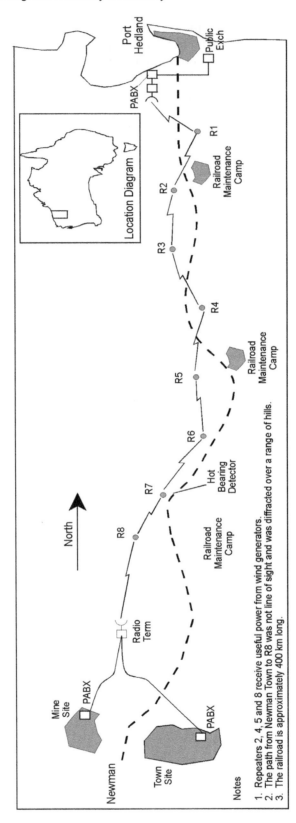

Figure 8.3
Railroad communications system

Appendix A

Glossary of terms

ABM   *asynchronous balanced mode*

ACE   *asynchronous communications element* – similar to UART

ACK   *acknowledge* (ASCII – control F)

Active filter   active circuit devices (usually amplifiers), with passive circuit elements (resistors and capacitors) and which have characteristics that more closely match ideal filters than do passive filters

Active passive device device capable of supplying the current for the loop (active) or one that must draw its power from connected equipment (passive)

Address   a normally unique designator for location of data or the identity of a peripheral device, which allows each device on a single communications line to respond to its own message

AFC   *automatic frequency control* – the circuit in a radio receiver that keeps the carrier frequency centered in the passband of the filters and demodulators automatically

AGC   *automatic gain control* – the circuit in a radio that keeps the carrier gain at the proper level automatically

Algorithm   normally used as a basis for writing a computer program. This is a set of rules with a finite number of steps for solving a problem

Alias frequency   a false lower frequency component that appears in data reconstructed from original data acquired at an insufficient sampling rate (which is less than two (2) times the maximum frequency of the original data)

ALU *arithmetic logic unit*

Amplitude flatness a measure of how close to constant the gain of a circuit remains over a range of frequencies

Amplitude modulation a modulation technique (also referred to as AM or ASK) used to allow data to be transmitted across an analog network, such as a switched telephone network. The amplitude of a single (carrier) frequency is varied or modulated between two levels one for binary 0 and one for binary 1

Analog a continuous real time phenomena where the information values are represented in a variable and continuous waveform

ANSI *American National Standards Institute* – the principal standards development body in the USA

Apogee the point in an elliptical orbit that is farthest from earth

Application layer the highest layer of the seven-layer ISO/OSI reference model structure, which contains all user or application programs

Arithmetic logic unit the element(s) in a processing system that perform(s) the mathematical functions such as addition, subtraction, multiplication, division, inversion, AND, OR, NAND and NOR

ARQ *automatic request for transmission* – a request by the receiver for the transmitter to retransmit a block or frame because of errors detected in the originally received message

AS *Australian Standard*

ASCII *American Standard Code for Information Interchange* – a universal standard for encoding alphanumeric characters into 7 or 8 binary bits, drawn up by ANSI to ensure compatibility between different computer systems

ASIC *application specific integrated circuit*

ASK *amplitude shift keying – see* Amplitude modulation

Asynchronous communications where characters can be transmitted at an arbitrary unsynchronized point in time and where the time intervals between transmitted characters may be of varying lengths – communication is controlled by start and stop bits at the beginning and end of each character

Attenuation the decrease in the magnitude of strength (or power) of a signal – in cables, generally expressed in dB per unit length

Attenuator a passive network that decreases the amplitude of a signal (without introducing any undesirable characteristics to the signals such as distortion)

Auto tracking antenna	a receiving antenna that moves in synchronism with the transmitting device, which is moving (such as a vehicle being telemetered)
AWG	*American wire gauge*
Balanced circuit	a circuit so arranged that the impressed voltages on each conductor of the pair are equal in magnitude but opposite in polarity with respect to ground
Band pass filter	a filter that allows only a fixed range of frequencies to pass through – all other frequencies outside this range (or band) are sharply reduced in magnitude
Band reject	a circuit that rejects a defined frequency band of signals while passing all signals outside this frequency range (both lower than and higher than)
Bandwidth	the range of frequencies available expressed as the difference between the highest and lowest frequencies is expressed in Hertz (or cycles per second)
Base address	a memory address that serves as the reference point – all other points are located by offsetting in relation to the base address
Base band	*base band* operation is the direct transmission of data over a transmission medium without the prior modulation on a high frequency carrier band
Base loading	an inductance situated near the bottom end of a vertical antenna to modify the electrical length – this aids in impedance matching
Baud	unit of signaling speed derived from the number of events per second (normally bits per second) – however if each event has more than one bit associated with it the baud rate and bits per second are not equal
Baudot	data transmission code in which five bits represents one character – sixty four (64) alphanumeric characters can be represented; this code is used in many teleprinter systems with one start bit and 1.42 stop bits added
BCC	*block check character* – error checking scheme with one check character; a good example being block sum check
BCD	*binary coded decimal* – a code used for representing decimal digits in a binary code
BEL	*bell* (ASCII for control G)
Bell 212	an AT&T specification of full duplex, asynchronous or synchronous 1200 baud data transmission for use on the public telephone network
BERT/BLERT	*bit error rate/block error rate testing* – an error checking technique that compares a received data pattern with a known transmitted data pattern to determine transmission line quality

Bifilar   two conducting elements used in parallel (such as two parallel wires wound on a coil form)

BIOS   *basic input/output system*

Bipolar   a signal range that includes both positive and negative values

BIT (binary digit) derived from 'BInary digiT', a one or zero condition in the binary system

Bit stuffing   *bit stuffing with zero bit insertion* – a technique used to allow pure binary data to be transmitted on a synchronous transmission line (each message block (frame) is encapsulated between two flags, which are special bit sequences. Then if the message data contains a possibly similar sequence, an additional (zero) bit is inserted into the data stream by the sender, and is subsequently removed by the receiving device. The transmission method is then said to be data transparent)

Bits per sec (BPS) unit of data transmission rate

Block sum check   this is used for the detection of errors when data is being transmitted. (It comprises a set of binary digits (bits), which are the Modulo 2 sum of the individual characters or octets in a frame (block) or message)

Bridge   a device to connect similar sub-networks without its own network address – used mostly to reduce the network load

Broad band   a communications channel that has greater bandwidth than a voice grade line and is potentially capable of greater transmission rates (*Opposite of base band.* In wide band operation the data to be transmitted are first modulated on a high frequency carrier signal. They can then be simultaneously transmitted with other data modulated on a different carrier signal on the same transmission medium)

Broadcast   a message on a bus intended for all devices which requires no reply

BS   *backspace* (ASCII control H)

BS   *British Standard*

BSC   *bisynchronous transmission* – a byte or character oriented communication protocol that has become the industry standard – created by IBM. (It uses a defined set of control characters for synchronized transmission of binary coded data between stations in a data communications system)

Buffer   an intermediate temporary storage device used to compensate for a difference in data rate and data flow between two devices (also called a spooler for interfacing a computer and a printer)

Burst mode   a high-speed data transfer in which the address of the data is sent followed by back-to-back data words while a physical signal is asserted

Bus	a data path shared by many devices with one or more conductors for transmitting signals, data or power
Byte	a term referring to eight associated bits of information; sometimes called a 'character'
Capacitance	storage of electrically separated charges between two plates having different potential. (The value is proportional to the surface area of the plates and inversely proportional to the distance between them)
Capacitance (mutual)	the capacitance between two conductors with all other conductors, including shield, short-circuited to the ground
Cascade	two or more electrical circuits in which the output of one is fed into the input of the next one
Cassegrain antenna	parabolic antenna that has a hyperbolic passive reflector situated at the focus of the parabola
CCITT	*Consultative Committee International Telegraph and Telephone* – an international association that sets worldwide standards (e.g. V.21, V.22, V.22bis). (Now referred to as the International Telecommunications Union (ITU))
Cellular polyethylene	expanded or 'foam' polyethylene consisting of individual closed cells suspended in a polyethylene medium
Channel selector	in an FM discriminator the plug-in module, which causes the device to select one of the channels and demodulate the subcarrier to recover data.
Character	letter, numeral, punctuation, control figure or any other symbol contained in a message
Characteristic impedance	the impedance that, when connected to the output terminals of a transmission line of any length, makes the line appear infinitely long. (The ratio of voltage to current at every point along a transmission line on which there are no standing waves)
Clock	the source(s) of timing signals for sequencing electronic events e.g. synchronous data transfer
Closed loop	a signal path that has a forward route for the signal, a feedback network for the signal and a summing point
CMRR	*common mode rejection ratio*
CMV	*common mode voltage*
CNR	*carrier to noise ratio* – an indication of the quality of the modulated signal

Common carrier   a private data communications utility company that furnishes communications services to the general public

Common mode signal   the common voltage to the two parts of a differential signal applied to a balanced circuit

Commutator   a device used to effect time-division multiplexing by repetitive sequential switching

Composite link   the line or circuit connecting a pair of multiplexers or concentrators; the circuit carrying multiplexed data

Conical scan antenna   an automatic tracking antenna system in which the beam is steered in a circular path so that it forms a cone

Contention   the facility provided by the dial network or a data PABX, which allows multiple terminals to compete on a first come, first served basis, for a smaller number of computer posts

Correlator   a device, which compares two signals and indicates the similarity between the two signals

CPU   *central processing unit*

CR   *carriage return* (ASCII control M)

CRC   *cyclic redundancy check* – an error-checking mechanism using a polynomial algorithm based on the content of a message frame at the transmitter and included in a field appended to the frame (at the receiver, it is then compared with the result of the calculation that is performed by the receiver. Also referred to as CRC-16)

Cross talk   a situation where a signal from a communications channel interferes with an associated channel's signals

Crossed planning   wiring configuration that allows two DTE or DCE devices to communicate. (Essentially it involves connecting pin 2 to pin 3 of the two devices)

Crossover   in communications, a conductor that runs through the cable and connects to a different pin number at each end

CSMA/CD   *carrier sense multiple access/collision detection* – when two stations transmit at the same time on a local area network, they both cease transmission and signal that a collision has occurred (each then tries again after waiting for a predetermined time period)

Current loop   communication method that allows data to be transmitted over a longer distance with a higher noise immunity level than with the standard EIA-232-C voltage method – a mark (a binary 1) is represented by current of 20 mA and a space (or binary 0) is represented by the absence of current

Data integrity   a performance measure based on the rate of undetected errors

Data link layer   this corresponds to layer 2 of the ISO reference model for open systems' interconnection – it is concerned with the reliable transfer of data (no residual transmission errors) across the data link being used

Data reduction   the process of analyzing a large quantity of data in order to extract some statistical summary of the underlying parameters

Datagram   a type of service offered on a packet-switched data network – a datagram is a self-contained packet of information that is sent through the network with minimum protocol overheads

dBi   a unit that is used to represent the gain of an antenna compared to the gain of an isotropic radiator

dBm   a signal level that is compared to a 1 mW reference

dBmV   a signal amplitude that is compared to a 1 mV reference

dBW   a signal amplitude that is compared to a 1 watt reference

DCE   *data communications equipment* – devices that provide the functions required to establish, maintain and terminate a data transmission connection (normally it refers to a modem)

Decibel (dB)   a logarithmic measure of the ratio of two signal levels where $dB = 20\log_{10}V1/V2$ or where $dB = 10\log_{10}P1/P2$ and where V refers to Voltage or P refers to Power. (*Note that it has no units of measure*)

Decoder   a device that converts a combination of signals into a single signal representing that combination

Decommutator   equipment for the demultiplexing of commutated signals

Default   a value or setup condition assigned, which is automatically assumed for the system unless otherwise explicitly specified

Delay distortion   distortion of a signal caused by the frequency components making up the signal having different propagation velocities across a transmission medium

DES   *data encryption standard*

Dielectric constant (E)   the ratio of the capacitance using the material in question as the dielectric, to the capacitance resulting when the material is replaced by air

Digital   a signal, which has definite states (normally two)

DIN   *Deutsches Institut Für Normierung*

DIP   acronym for dual in line package referring to integrated circuits and switches

Diplexing   a device used to allow simultaneous reception or transmission of two signals on a common antenna

Direct memory access   a technique of transferring data between the computer memory and a device on the computer bus without the intervention of the microprocessor. Also abbreviated to DMA

Discriminator   hardware device to demodulate a frequency modulated carrier or subcarrier to produce analog data

Dish   a concave antenna reflector for use at VHF or higher frequencies

Dish antenna   an antenna in which a parabolic dish acts a reflector to increase the gain of the antenna

Diversity reception   two or more radio receivers connected to different antennas to improve signal quality by using two different radio signals to transfer the information

DLE   *data link escape* (ASCII character)

DNA   *distributed network architecture*

Doppler   the change in observed frequency of a signal caused by the emitting device moving with respect to the observing device

Downlink   the path from a satellite to an earth station

DPI   *dots per inch*

DPLL   *digital phase locked loop*

DR   *dynamic range* – the ratio of the full-scale range (FSR) of a data converter to the smallest difference it can resolve. $DR = 2n$ where n is the resolution in bits

Drift   a (normally gradual) change in a component's characteristics over time

Driver software   a program that acts as the interface between a higher level coding structure and the lower level hardware/firmware component of a computer

DSP   *digital signal processing*

DSR   *data set ready* – an EIA-232 modem interface control signal, which indicates that the terminal is ready for transmission

DTE	*data terminal equipment* – devices acting as data source, data sink, or both
Duplex	the ability to send and receive data simultaneously over the same communications line
Dynamic range	the difference in decibels between the overload or maximum and minimum discernible signal level in a system
EBCDIC	*extended binary coded decimal interchange code* – an eight bit character code used primarily in IBM equipment. (The code allows for 256 different bit patterns)
EDAC	*error detection and correction*
EIA	*Electronic Industries Association* – a standards organization in the USA specializing in the electrical and functional characteristics of interface equipment
EIA-232-C	interface between DTE and DCE, employing serial binary data exchange (typical maximum specifications are 15 m at 19 200 Baud)
EIA-422	interface between DTE and DCE employing the electrical characteristics of balanced voltage interface circuits
EIA-423	interface between DTE and DCE, employing the electrical characteristics of unbalanced voltage digital interface circuits
EIA-449	general-purpose 37 pin and 9 pin interface for DCE and DTE employing serial binary interchange
EIA-485	the recommended standard of the EIA that specifies the electrical characteristics of drivers and receivers for use in balanced digital multipoint systems
EIRP	*effective isotropic radiated power* – the effective power radiated from a transmitting antenna when an isotropic radiator is used to determine the gain of the antenna
EISA	*enhanced industry standard architecture*
EMI/RFI	*electromagnetic interference/radio frequency interference* – 'background noise' that could modify or destroy data transmission
EMS	*expanded memory specification*
Emulation	the imitation of a computer system performed by a combination of hardware and software that allows programs to run between incompatible systems
Enabling	the activation of a function of a device by a defined signal

Encoder   a circuit, which changes a given signal into a coded combination for purposes of optimum transmission of the signal

ENQ   *enquiry* (ASCII control E)

EOT   *end of transmission* (ASCII control D)

EPROM   *erasable programmable read only memory* – non-volatile semiconductor memory that is erasable in an ultra violet light and reprogrammable

Equalizer   the device, which compensates for the unequal gain characteristic of the signal received

Error rate   the ratio of the average number of bits that will be corrupted to the total number of bits that are transmitted for a data link or system

ESC   *escape* (ASCII character)

ESD   *electrostatic discharge*

Ethernet   name of a widely used LAN, based on the CSMA/CD bus access method (IEEE 802.3). (Ethernet is the basis of the TOP bus topology)

ETX   *end of text* (ASCII control C)

Even parity   a data verification method normally implemented in hardware in which each character must have an even number of 'ON' bits

Farad   unit of capacitance whereby a charge of one coulomb produces a one-volt potential difference

Faraday rotation rotation of the plane of polarization of an electromagnetic wave when traveling through a magnetic field

FCC   *Federal Communications Commission*

FCS   *frame check sequence* – a general term given to the additional bits appended to a transmitted frame or message by the source to enable the receiver to detect possible transmission errors

FDM   *frequency division multiplexer* – a device that divides the available transmission frequency range in narrower bands, each of which is used for a separate channel

FDM   *frequency division multiplexing* – the combining of one or more signals (each occupying a defined non overlapping frequency band) into one signal

Feedback   a part of the output signal being fed back to the input of the amplifier circuit

FIFO   *first in, first out*

Filled cable	a telephone cable construction in which the cable core is filled with a material that will prevent moisture from entering or passing along the cable
FIP	*factory instrumentation protocol*
Firmware	a computer program or software stored permanently in PROM or ROM or semi-permanently in EPROM
Flame retardancy	the ability of a material not to propagate flame once the flame source is removed
Floating	an electrical circuit that is above the earth potential
Flow control	the procedure for regulating the flow of data between two devices preventing the loss of data once a device's buffer has reached its capacity
Frame	the unit of information transferred across a data link. (Typically, there are control frames for link management and information frames for the transfer of message data)
Frequency	refers to the number of cycles per second
Frequency domain	the displaying of electrical quantities versus frequency
Frequency modulation	a modulation technique (abbreviated to FM) used to allow data to be transmitted across an analog network where the frequency is varied between two levels – one for binary '0' and one for binary '1'. (Also known as frequency shift keying (or FSK))
Full duplex	simultaneous two-way independent transmission in both directions (4 wire). (See Duplex)
G	*Giga* (metric system prefix – 10^9)
Gain of antenna	the difference in signal strengths between a given antenna and a reference isotropic antenna
Gateway	a device to connect two different networks that translates the different protocols
Geostationary	a special earth orbit that allows a satellite to remain in a fixed position above the equator
Geosynchronous	any earth orbit in which the time required for one revolution of a satellite is an integral portion of a sidereal day.
GPIB	*general purpose interface bus* – an interface standard used for parallel data communication, usually used for controlling electronic instruments from a computer (also designated IEEE-488 standard)

Ground	an electrically neutral circuit having the same potential as the earth. (A reference point for an electrical system also intended for safety purposes)
Half duplex	transmissions in either direction, but not simultaneously
Half power point	the point in a power vs frequency curve which is half the power level of the peak power (also called the 3 dB point)
Hamming distance	a measure of the effectiveness of error checking (the higher the Hamming distance (HD) index, the safer is the data transmission)
Handshaking	exchange of predetermined signals between two devices establishing a connection
Harmonic distortion	distortion caused by the presence of harmonics in the desired signal
HDLC	*high level data link control* – the international standard communication protocol defined by ISO to control the exchange of data across either a point-to-point data link or a multidrop data link
Hertz (Hz)	a term replacing cycles per second as a unit of frequency
Hex	*hexadecimal*
HF	*high frequency*
High pass	generally referring to filters, which allow signals above a specified frequency to pass but attenuate signals below this, specified frequency
Horn	a moderate-gain wide-beamwidth antenna
Host	this is normally a computer belonging to a user that contains (hosts) the communication hardware and software necessary to connect the computer to a data communications network
IA5	*international alphabet number 5*
IEC	*International Electrotechnical Commission*
IEE	*Institution of Electrical Engineers*
IEEE	*Institute of Electrical and Electronic Engineers* – an American based international professional society that issues its own standards and is a member of ANSI and ISO
Impedance	the total opposition that a circuit offers to the flow of alternating current or any other varying current at a particular frequency (it is a combination of resistance R and reactance X, measured in ohms)

Inductance the property of a circuit or circuit element that opposes a change in current flow, thus causing current changes to lag behind voltage changes (it is measured in henrys)

Insulation resistance (IR) that resistance offered by an insulation to an impressed DC voltage, tending to produce a leakage current though the insulation

Interface a shared boundary defined by common physical interconnection characteristics, signal characteristics and measurement of interchanged signals

Interrupt handler the section of the program that performs the necessary operation to service an interrupt when it occurs

Interrupt an external event indicating that the CPU should suspend its current task to service a designated activity

I/O address a method that allows the CPU to distinguish between different boards in a system – all boards must have different addresses

IP *Internet protocol*

ISA *industry standard architecture* (for IBM personal computers)

ISB *intrinsically safe barrier*

ISDN *integrated services digital network* – the new generation of world-wide telecommunication's network that utilizes digital techniques for both transmission and switching – it supports both voice and data communications

ISO *International Standards Organization*

Isotropic antenna a reference antenna that radiates energy in all directions from a point source

ISR *interrupt service routine* (*see* Interrupt handler)

ITU *International Telecommunications Union*

Jumper a wire connecting one or more pins on the one end of a cable only

k (kilo) this is 2^{10} or 1024 in computer terminology, e.g. 1 kB = 1024 bytes

LAN *local area network* – a data communications system confined to a limited geographic area typically about 10 kms with moderate to high data rates (100 kbps to 100 Mbps). (Some type of switching technology is used, but common carrier circuits are not used)

LCD *liquid crystal display* – a low power display system used on many laptops and other digital equipment

LDM *limited distance modem* – a signal converter that conditions and boosts a digital signal so that it may be transmitted further than a standard EIA-232 signal

Leased (or private) line a private telephone line without inter-exchange switching arrangements

LED *light emitting diode* – a semi-conductor light source that emits visible light or infrared radiation

LF *line feed* (ASCII control J)

Line driver a signal converter that conditions a signal to ensure reliable transmission over an extended distance

Line turnaround the reversing of transmission direction from transmitter to receiver or vice versa when a half duplex circuit is used

Linearity a relationship where the output is directly proportional to the input

Link layer layer 2 of the ISO/OSI reference model (also known as the data link layer)

Listener a device on the GPIB bus that receives information from the bus

LLC *logical link control* (IEEE 802)

Loaded line a telephone line equipped with loading coils to add inductance in order to minimize amplitude distortion

Long wire a horizontal wire antenna that is one wavelength or greater in size

Loop resistance the measured resistance of two conductors forming a circuit

Loopback type of diagnostic test in which the transmitted signal is returned on the sending device after passing through all, or a portion of, a data communication link or network – a loopback test permits the comparison of a returned signal with the transmitted signal

Low pass generally referring to filters which allow signals below a specified frequency to pass but not attenuated signals above this specified frequency

m *Meter* (Metric system unit for length)

M *Mega* (Metric system prefix for 10^6)

MAC *Media Access Control* (IEEE 802)

Manchester encoding digital technique (specified for the IEEE-802.3 Ethernet baseband network standard) in which each bit period is divided into two complementary halves – a negative to positive voltage transition in the middle of the bit period designates a binary '1', whilst a positive to negative transition represents a '0' (the encoding technique also allows the receiving device to recover the transmitted clock from the incoming data stream (self clocking))

Mark   this is equivalent to a binary 1

Master oscillator   the primary oscillator for controlling a transmitter or receiver frequency (the various types are: variable frequency oscillator (VFO); variable crystal oscillator (VXO); permeability tuned oscillator (PTO); phase locked loop (PLL); linear master oscillator (LMO) or frequency synthesizer)

Master/slave   bus access method whereby the right to transmit is assigned to one device only, the master, and all the other devices, the slaves may only transmit when requested

Microwave   AC signals having frequencies of 1 GHz or more

MIPS   *million instructions per second*

Modem   *MODulator/DEModulator* – a device used to convert serial digital data from a transmitting terminal to a signal suitable for transmission over a telephone channel or to reconvert the transmitted signal to serial digital data for the receiving terminal

Modem eliminator   a device used to connect a local terminal and a computer port in lieu of the pair of modems to which they would ordinarily connect, allow DTE to DTE data and control signal connections otherwise not easily achieved by standard cables or connections

Modulation index   the ratio of the frequency deviation of the modulated wave to the frequency of the modulating signal

MOS   *metal oxide semiconductor*

MOV   *metal oxide varistor*

MTBF   *mean time between failures*

MTTR   *mean time to repair*

Multidrop   a single communication line or bus used to connect three or more points

Multiplexer (MUX)   a device used for division of a communication link into two or more channels either by using frequency division or time division

NAK   *negative acknowledge* (ASCII control U)

Narrowband   a device that can only operate over a narrow band of frequencies

Network   an interconnected group of nodes or stations

Network architecture　a set of design principles including the organization of functions and the description of data formats and procedures used as the basis for the design and implementation of a network (ISO)

Network layer　layer 3 in the ISO/OSI reference model, the logical network entity that services the transport layer responsible for ensuring that data passed to it from the transport layer is routed and delivered throughout the network

Network topology　the physical and logical relationship of nodes in a network; the schematic arrangement of the links and nodes of a network typically in the form of a star, ring, tree or bus topology

NMRR　*normal mode rejection ratio*

Node　a point of interconnection to a network

Noise　a term given to the extraneous electrical signals that may be generated or picked up in a transmission line. (If the noise signal is large compared with the data carrying signal, the latter may be corrupted resulting in transmission errors)

Non-linearity　a type of error in which the output from a device does not relate to the input in a linear manner

NRZ　*non return to zero* – pulses in alternating directions for successive 1 bits but no change from existing signal voltage for 0 bits

NRZI　*non return to zero inverted*

Null modem　a device that connects two DTE devices directly by emulating the physical connections of a DCE device

Nyquist sampling theorem　in order to recover all the information about a specified signal it must be sampled at least at twice the maximum frequency component of the specified signal

Ohm (Ω)　unit of resistance such that a constant current of one ampere produces a potential difference of one volt across a conductor

Optical isolation　two networks with no electrical continuity in their connection because an optoelectronic transmitter and receiver has been used

OSI　*open systems interconnection*

Packet　a group of bits (including data and call control signals) transmitted as a whole on a packet switching network. (Usually smaller than a transmission block)

PAD　*packet access device* – an interface between a terminal or computer and a packet switching network

Parallel transmission   the transmission model where a number of bits is sent simultaneously over separate parallel lines (usually unidirectional such as the Centronics interface for a printer)

Parametric amplifier   an inverting parametric device for amplifying a signal without frequency translation from input to output

Parasitic   undesirable electrical parameter in a circuit such as oscillations or capacitance

Parity bit   a bit that is set to a '0' or '1' to ensure that the total number of 1 bits in the data field is even or odd

Parity check   the addition of non information bits that make up a transmission block to ensure that the total number of bits is always even (even parity) or odd (odd parity) – used to detect transmission errors but rapidly losing popularity because of its weakness in detecting errors

Passive filter   a circuit using only passive electronic components such as resistors, capacitors and inductors

Path loss   the signal loss between transmitting and receiving antennas

PBX   *private branch exchange*

PCM   *pulse code modulation* – the sampling of a signal and encoding the amplitude of each sample into a series of uniform pulses

PEP   *peak envelope power* – maximum amplitude that can be achieved with any combination of signals

Perigee   the point in an elliptical orbit that is closest to earth

Peripherals   the input/output and data storage devices attached to a computer e.g. disk drives, printers, keyboards, display, communication boards, etc

Phase modulation   the sine wave or carrier has its phase charged in accordance with the information to be transmitted

Phase shift keying   a modulation technique (also referred to as PSK) used to convert binary data into an analog form comprising a single sinusoidal frequency signal whose phase varies according to the data being transmitted

Physical layer   layer 1 of the ISO/OSI reference model, concerned with the electrical and mechanical specifications of the network termination equipment

PLC   *programmable logic controller*

PLL   *phase locked loop*

Point-to-point   a connection between only two items of equipment

Polar orbit the path followed when the orbital plane includes the north and south poles

Polarization the direction of an electric field radiated from an antenna

Polling a means of controlling devices on a multipoint line. (A controller will query devices for a response)

Polyethylene a family of insulators derived from the polymerization of ethylene gas and characterized by outstanding electrical properties, including high IR, low dielectric constant, and low dielectric loss across the frequency spectrum

Polyvinyl chloride (PVC) a general purpose family of insulations whose basic constituent is polyvinyl chloride or its copolymer with vinyl acetate – plasticizers, stabilizers, pigments and fillers are added to improve mechanical and/or electrical properties of this material

Port a place of access to a device or network, used for input/output of digital and analog signals

Presentation layer layer 6 of the ISO/OSI reference model, concerned with negotiation of a suitable transfer syntax for use during an application. (If this is different from the local syntax, the translation to/from this syntax)

Protocol a formal set of conventions governing the formatting, control procedures and relative timing of message exchange between two communicating systems

PSDN *public switched data network* – any switching data communications system, such as Telex and public telephone networks, which provides circuit switching to many customers

PSTN *public switched telephone network* – this is the term used to describe the (analog) public telephone network

PTT *Post, Telephone and Telecommunications Authority*

QAM *quadrature amplitude modulation*

QPSK *quadrature phase shift keying*

Quagi an antenna consisting of both full wavelength loops (quad) and Yagi elements

RAM *random access memory* – semiconductor read/write volatile memory (data is lost if the power is turned off)

Reactance the opposition offered to the flow of alternating current by inductance or capacitance of a component or circuit

Repeater an amplifier, which regenerates the signal and thus expands the network

Resistance the ratio of voltage to electrical current for a given circuit measured in ohms

Response time	the elapsed time between the generation of the last character of a message at a terminal and the receipt of the first character of the reply (it includes terminal delay and network delay)
RF	*radio frequency*
RFI	*radio frequency interference*
Ring	network topology commonly used for interconnection of communities of digital devices distributed over a localized area, e.g. a factory or office block (each device is connected to its nearest neighbors until all the devices are connected in a closed loop or ring – data is transmitted in one direction only – as each message circulates around the ring, it is read by each device connected in the ring)
Ringing	an undesirable oscillation or pulsating current
Rise time	the time required for a waveform to reach a specified value from some smaller value
RMS	*root mean square*
ROM	*read only memory* – computer memory in which data can be routinely read but written to only once using special means, when the ROM is manufactured (a ROM is used for storing data or programs on a permanent basis)
Router	a linking device between network segments which may differ in Layers 1, 2a and 2b of the ISO/OSI Reference Model
RS	*recommended standard* (e.g. RS-232-C) (newer designations use the prefix EIA (e.g. EIA-RS-232-C or just EIA-232-C))
RTU	*remote terminal unit* – terminal unit situated remotely from the main control system
R/W	*read/write*
SAA	*Standards Association of Australia*
SAP	*service access point*
SDLC	*synchronous data link control* – IBM standard protocol superseding the bisynchronous standard
Selectivity	a measure of the performance of a circuit in distinguishing the desired signal from those at other frequencies
Serial transmission	the most common transmission mode in which information bits are sent sequentially on a single data channel

Session layer   layer 5 of the ISO/OSI Reference Model, concerned with the establishment of a logical connection between two application entities and with controlling the dialog (message exchange) between them

Short haul modem   a signal converter, which conditions a digital signal for transmission over DC continuous private line metallic circuits, without interfering with adjacent pairs of wires in the same telephone cables

Sidebands   the frequency components which are generated when a carrier is frequency-modulated

Sidereal day   the period of an earth's rotation with respect to the stars

Signal to noise ratio   the ratio of signal strength to the level of noise

Simplex transmissions data transmission in one direction only

Slew rate   this is defined as the rate at which the voltage changes from one value to another

SNA   *systems network architecture*

SNR   *signal to noise ratio*

SOH   *start of header* (ASCII Control-A)

Space   *absence of signal* – this is equivalent to a binary 0

Spark test   a test designed to locate imperfections (usually pin-holes) in the insulation of a wire or cable by application of a voltage for a very short period of time while the wire is being drawn through the electrode field

Spectral purity   the relative quality of a signal measured by the absence of harmonics, spurious signals and noise

Standing wave ratio   the ratio of the maximum-to-minimum voltage (or current) on a transmission line at least a quarter-wavelength long (VSWR refers to voltage standing wave ratio)

Star   a type of network topology in which there is a central node that performs all switching (and hence routing) functions

Statistical multiplexer   a device used to enable a number of lower bit rate devices normally situated in the same location, to share a single, higher bit rate transmission line (the devices usually have human operators and hence data are transmitted on the shared line on a statistical basis rather than, as is the case with a basic multiplexer, on a pre-allocated basis – it thus endeavors to exploit the fact that each device operates at a much lower mean rate than its maximum rate)

STP	*shielded twisted pair*
Straight through pinning	EIA-232 and EIA-422 configuration that match DTE to DCE, pin for pin (pin 1 with pin 1, pin 2 with pin 2, etc)
STX	*start of text* (ASCII control B)
Subharmonic	a frequency that is an integral submultiple of a reference frequency
Switched line	a communication link for which the physical path may vary with each usage, such as the public telephone network
Synchronization	the co-ordination of the activities of several circuit elements
Synchronous transmission	transmission in which data bits are sent at a fixed rate, with the transmitter and receiver synchronized (synchronized transmission eliminates the need for start and stop bits)
Talker	a device on the GPIB bus that simply sends information onto the bus without actually controlling the bus
Tank	a circuit comprising inductance and capacitance, which can store electrical energy over a finite band of frequencies
TCP	*transmission control protocol*
TDM	*time division multiplexer* – a device that accepts multiple channels on a single transmission line by connecting terminals, one at a time, at regular intervals, interleaving bits (bit TDM) or characters (Character TDM) from each terminal
Telegram	in general a data block that is transmitted on the network (usually comprises address, information and check characters)
Temperature rating	the maximum, and minimum temperature at which an insulating material may be used in continuous operation without loss of its basic properties
TIA	Telecommunications Industry Association
Time division multiplexing	the process of transmitting multiple signals over a single channel by taking samples of each signal in a repetitive time sequenced fashion
Time sharing	a method of computer operation that allows several interactive terminals to use one computer
Time domain	the display of electrical quantities versus time

Token ring   collision free, deterministic bus access method as per IEEE 802.2 ring topology

TOP   *Technical Office Protocol* – a user association in USA, which is primarily concerned with open communications in offices

Topology   physical configuration of network nodes, e.g. bus, ring, star, tree

Transceiver   *transmitter/receiver* – Network access point for IEEE 803.2 networks; a combination of transmitter and receiver

Transient   an abrupt change in voltage of short duration

Transmission line   one or more conductors used to convey electrical energy from one point to another

Transport layer   layer 4 of the ISO/OSI reference model, concerned with providing a network independent reliable message interchange service to the application oriented layers (layers 5 through 7)

Trunk   a single circuit between two points, both of which are switching centers or individual distribution points (a trunk usually handles many channels simultaneously)

Twisted pair   a data transmission medium, consisting of two insulated copper wires twisted together (this improves its immunity to interference from nearby electrical sources that may corrupt the transmitted signal)

UART   *universal asynchronous receiver/transmitter* – an electronic circuit that translates the data format between a parallel representation, within a computer, and the serial method of transmitting data over a communications line

UHF   *ultra high frequency*

Upconverter   a device used to translate a modulated signal to a higher band of frequencies

Unbalanced circuit   a transmission line in which voltages, on the two conductors, is unequal with respect to ground e.g. a coaxial cable

Unloaded line   a line with no loaded coils that reduce line loss at audio frequencies

Uplink   the path from an earth station to a satellite

USRT   *universal synchronous receiver/transmitter* (See UART)

UTP   *unshielded twisted pair*

V.35   CCITT standard governing the transmission at 48 kbps over 60 to 108 kHz group band circuits

VCO	*voltage controlled oscillator* – uses variable DC applied to tuning diodes to change their junction capacitances (this results in the output frequency being dependent on the input voltage)
Velocity of propagation	the speed of an electrical signal down a length of cable compared to speed in free space expressed as a percentage
VHF	*very high frequency*
Volatile memory	an electronic storage medium that loses all data when power is removed
Voltage rating	the highest voltage that may be continuously applied to a wire in conformance with standards of specifications
VSD	*variable speed drive*
VT	*virtual terminal*
WAN	*wide area network*
Waveguide	a hollow conducting tube used to convey microwave energy
Word	the standard number of bits that a processor or memory manipulates at one time – typically, a word has 16 bits
X.21	CCITT standard governing interface between DTE and DCE devices for synchronous operation on public data networks
X.25	CCITT standard governing interface between DTE and DCE device for terminals operating in the packet mode on public data networks
X.25 pad	a device that permits communication between non X.25 devices and the devices in an X.25 network
X.3/X.28/X.29	a set of internationally agreed standard protocols defined to allow a character oriented device, such as a visual display terminal, to be connected to a packet switched data network
X-ON/X-OFF	*transmitter on/transmitter off* – control characters used for flow control, instructing a terminal to start transmission (X-ON or control S) and end transmission (X-OFF or control Q)

Appendix B

Path loss calculation formulae

The following section provides two spreadsheets, and their associated formula, for calculating path loss and availability figures for radio and microwave links. The formulae are provided with direct references to the spreadsheet cell numbers. The user can enter the relevant data into the spreadsheet, and then using the formulas, calculate the relative availability of the link.

It should be noted that there are a number of software packages available for carrying out radio path analysis that would make this process significantly easier.

B.1   Radio spreadsheet

	PARAMETER	SITE 1		SITE 2	
		TX	RX	TX	RX
1	Latitude				
2	Longitude				
3	Calculated distance elevation (km)				
4	Elevation (m)				
5	Antenna height (m)				
6	Path length				
7	Frequencies (MHz)				
8	Free space attenuation (dB)				
9	Antenna type				

	PARAMETER	SITE 1		SITE 2	
		TX	RX	TX	RX
10	Antenna gain				
11	dBd				
12	dBi				
13	Feeder type				
14	Feeder length (m)				
15	Feeder attenuation (dB/100 m)				
16	Feeder losses (dB)				
17	Connector losses (dB)				
18	Diplexer loss (dB)				
19	TX multicoupler loss (dB)				
20	Isolator loss (dB)				
21	Receiver filter loss (dB)				
22	Receiver pre-amp gain (dB)				
23	Diffraction losses (dB)				
24	Other losses (dB)				
25	Total losses · Site 1 TX to site 2 RX · Site 2 TX to site 1 RX				
26	Transmit power				
27	· Watts				
28	· dBm				
29	Receiver sensitivity				
30	· dBm				
31	· μV				
32	12 dB SINAD sensitivity				
33	· dBm				
34	· μV				
35	Fade margin for squelch				

	PARAMETER	SITE 1		SITE 2	
		TX	**RX**	**TX**	**RX**
36	· Site 1 to site 2				
37	· Site 2 to site 1				
38	Fade margin for 12 dB SINAD (dB)				
39	· Site 1 to site 2				
40	· Site 2 to site 1				
41	General terrain type	Average ↓	Land ↓		
42	General weather type	Temp ↓			
43	Availability squelch (%)				
44	· Site 1 to site 2				
45	· Site 2 to site 1				
46	Availability 12 dB SINAD (%)				
47	· Site 1 to site 2				
48	· Site 2 to site 1				

B.2 Radio spreadsheet calculations

FOR (7)

Free space attenuation

$$AdB = 32.4 + 20\,Log_{10}\,F_{(MHz)} + 20\,Log_{10}\,D_{(kms)}$$

FOR (9)

$$dBd = 2.15 \times dBi$$

FOR (22)

Total Losses = Site 1 TX to site 2 TX =

– free space attenuation

+ antenna gain in dBi for site 1 TX

+ antenna gain in dBi for site 2 RX

– feeder loss site 1 TX

– feeder loss site 2 RX

– connector losses site 1 TX

– connector losses site 2 RX

– diplexer loss site 1 TX

– diplexcr loss site 2 RX

– multicoupler loss site 1 TX

– isolator loss site 1 TX

– receiver filter losses site 2 RX

+ receiver pre-amp gain site 2 RX

– diffraction losses

– reflection losses

Total losses site 2 *TX to site* 1 *RX =*

– free space attenuation

+ antenna gain in dBi for site 2 TX

+ antenna gain in dBi for site 1 RX

– feeder loss site 2 TX

– feeder loss site 1 RX

– connector losses site 2 TX

– connector losses site 1 RX

– diplexer loss site 2 TX

– diplexer loss site 1 RX

-    multicoupler loss site 2 TX

-    isolator loss site 2 TX

-    receiver filter losses site 1 RX

+   receiver pre-amp gain site 1 RX

-    diffraction losses

-    reflection losses

FOR (23)

$$Transmit\ power\ dBm = 10\ Log_{10}\left(\frac{Pwatts}{10^{-3}}\right)$$

$$\therefore Pwatts = 10^{\frac{TxPdBm}{10}} \times 10^{-3}$$

FOR (24) AND (25)

Receiver sensitivity and SINAD

$$Power\ dBm = 10\ Log_{10}\ \frac{\frac{V^2}{50}}{10^{-3}}$$

$$\therefore V = \sqrt{\frac{10^{\frac{PdBm}{10}} \times 50}{10^{-3}}}$$

Assuming 50 Ω antenna impedance

FOR (26)

Fade margin in site 1 *to site* 2 =

+   transmit power site 1 TX (in dBm)

+   total losses site 1 TX to site 2 RX

-    receiver sensitivity site 2

Fade margin in site 2 *to site* 1 =

+   transmit power site 2 TX

+   total losses site 2 TX to site 1 RX

–   receiver sensitivity site 1

FOR (27)

As for (26) except use 12 dB SINAD sensitivity

FOR (28 & 29)

Rayleigh availability for a multipath environment.

$$Availability = 100 \times e^{-\left(10^{\left(\frac{-Ft}{10} \right)} \right)}$$

$$Ft = \text{Fade margin}$$

B.3   Reflection analysis

Always assume worst case, i.e. that the wave will be totally reflected. What is flat, dry grassland today is a mirror pond after it rains.

NOTE:

- At the point of reflection, the wave changes 180° in phase.
- The first Fresnel zone in phase addition at the receiving antenna.
- The second Fresnel zone is phase subtraction at the receiving antenna (see attached diagram).
- Calculate anticipated reflection areas according to formula.

Below 0.6 F1 is grazing incidence, which is degrading in a linear fashion, therefore for grazing:

$$y = 33.33x - 20$$

Where x = fraction of Fresnel zone
y = dB attenuation

B.4   Diffraction losses

Most will be *knife edge*.
For example:

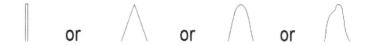

Figure B.1
Some will be smooth edge.
For example:

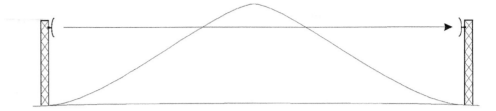

Figure B.2

For knife edge:
Firstly calculate v

$$v = -h_p \sqrt{\frac{2 \times F}{3 \times 10^8} \times \frac{(d_1 \times d_2)}{(d_1 \times d_2)}}$$

F in Hz
d in meters
h_p in meters

In this case V is V_e

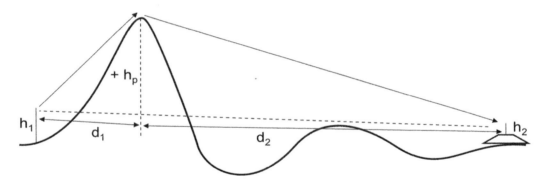

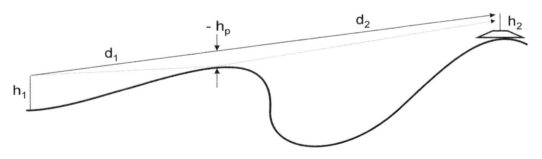

Figure B.3

In this case V is V_e

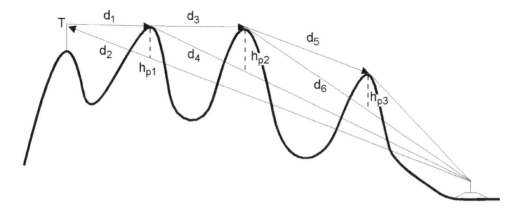

Figure B.4

For the various values of 'v' calculate the loss for each knife edge.

$$_0 L_R = 0\,\text{dB} \quad 1 \le v$$

$$_1 L_R = 20\,Log\,(\,0.5 + 0.62\,v\,) \quad 0 \le v \le 1$$

$$_2 L_R = 20\,Log\,(\,0.5\,e^{0.95v}\,) \quad -1 \le v \le 0$$

$$_3 L_R = 20\,Log\left(0.4 - \sqrt{0.1184 - (0.1v - 0.38)^2}\,\right) \quad -2.4 \le v < -1$$

$$_4 L_R = 20\,Log\left(-0.\frac{225}{v}\right) \quad v < -2.4$$

For multiple knife edges then with reference to Figure C.5, do a series addition of 3 knife edge losses, where $A = L_1 + L_2 + L_3.$

L_1 is for hp_1, d_1 and d_2
L_2 is for hp_2, d_3 and d_4
L_3 is for hp_3, d_5 and d_6

Calculating a new V in each case and then add them up.

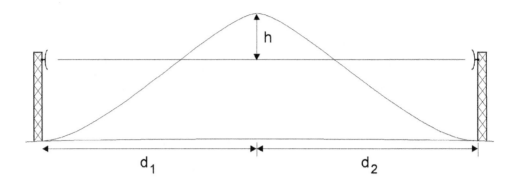

Figure B.5

For smooth earth:

$$Att\ dB = -\left(\dfrac{33.33\ h}{\sqrt[547.1]{\dfrac{d_1\,d_2}{F_{MHz}\,D}}}\right) - 20$$

$$F\ in\ MHz$$

$$D = d_1 + d_2$$

$$D, d_1, d_2\ in\ km$$

$$h\ in\ meters$$

B.5   Great circle distance

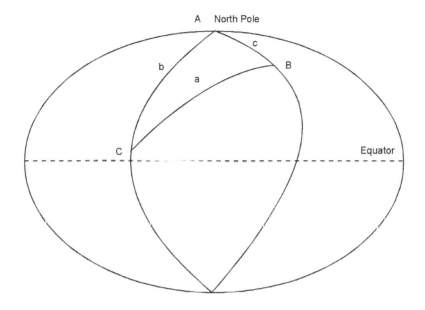

Figure B.6

For distance C to B

$$\cos a = \cos b \cos c + \sin b \sin c \cos A$$

B.6   Fresnel zone radius

$$F = {}_{0.55}\sqrt{\frac{n\, d_1\, d_2}{f\, (\ MHz\ ) \times D}}$$

F	$=$	Fresnel zone radius (in meters)
d_1	$=$	distance from site 1 to contour point (km)
d_2	$=$	distance from site 2 to contour point (km)
D	$=$	total length in km
F	$=$	frequency in (Mhz)
n	$=$	number Fresnel zone

B.7   Space diversity

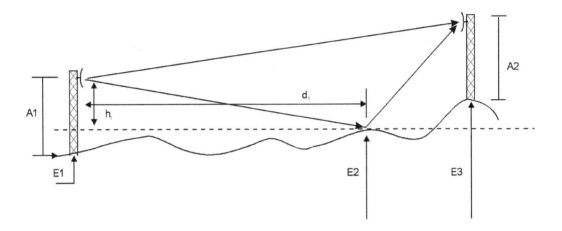

Figure B.7

E1, E2 & E3 are elevations
A1 & A2 are antenna heights

h_1 = (E1 + A1) − E2   (Without earth correction factors)
h_2 = (E3 + A2) − E2

Calculate the distance between the two antennas for space diversity.

$$\Delta h_t = h_1 - \frac{(d_1)^2}{17}$$

$$\Delta h_2 = \frac{127 \times D}{FGHz \times h_1} \qquad (K = \infty)$$

$$\Delta h_2 = \frac{75 \times D}{F_{GH} \times h_1} \qquad (K = \frac{4}{3})$$

Choose the smaller of the Δh_2 results.

B.8   Microwave spreadsheet

	PARAMETER	SITE 1	SITE 2
1	Latitude		
2	Longitude		
3	Great circle distance (km)		
4	Antenna height (m)		
5	Elevation (m)		
6	Path length (km)		
7	Frequencies (GHz)		
8	Free space attenuation (dB)		
9	Antenna type		
10	Antenna gain		
11	· dBd		
12	· dBi		
13	Feeder type		
14	Feeder length (m)		
15	Feeder attenuation (dB/100 m)		
16	Feeder losses (dB)		
17	Connector losses (dB)		
18	Branching equipment loss (dB)		
19	Isolator loss (dB)		
20	Radome loss (dB)		

	PARAMETER	SITE 1	SITE 2
21	Diffraction losses (dB)		
22	Other losses		
23	Total Losses · Site 1 TX to site 2 RX · Site 2 TX to site 1 RX		
24	Transmit power		
25	· watts		
26	· dBm		
27	Receiver sensitivity		
28	· dBm		
29	· μV		
30	Fade margin		
31	· Site 1 to site 2		
32	· Site 2 to site 1		
33	General terrain type	Average ↓	Land ↓
34	General weather type	Temp ↓	
35	Availability		
36	· Site 1 to site 2		
37	· Site 2 to site 1		
38	Rain attenuation		
39	Rain rate (mm/hr)		
40	% yr > = rain rate		
41	Hours per year		
42	Attenuation/km		
43	Total rain attenuation		

	PARAMETER	SITE 1	SITE 2
44	Availability with rain attenuation		
45	· Site 1 to site 2 (%)		
46	· Site 2 to site 1 (%)		
47	Space diversity		
48	Antenna separation		
49	· Site 1 Rx (m)		
50	· Site 2 Rx (m)		
51	Diffraction loss for diversity antenna		
52	· Site 1 (dB)		
53	· Site 2 (dB)		
54	Availability with space diversity		
55	· Site 1 to site 2 (%)		
56	· Site 2 to site 1 (%)		
57	Frequency diversity		
58	Frequency separation		
59	· Site 1 Tx (GHz)		
60	· Site 2 Tx (GHz)		
61	Availability with frequency diversity		
62	· Site 1 to site 2 (%)		
63	· Site 2 to site 1 (%)		
64	Hybrid diversity		
65	Availability with hybrid diversity		
66	· Site 1 to site 2 (%)		
67	· Site 2 to site 1 (%)		

B.9 Microwave radio spreadsheet calculations

FOR (6)

Free space attenuation

$$AdB = 92.4 + 20 \, Log_{10} \, F_{(GHz)}$$

$$+ \, 20 \, Log_{10} \, D_{(km)}$$

FOR (22)

Total losses = site 1 *to site* 2 =

−   free space attenuation

+   antenna gain in dBi for site 1

+   antenna gain in dBi for site 2

−   feeder loss site 1

−   feeder loss site 2

−   connector losses site 1

−   connector losses site 2

−   branching equipment loss site 1

−   branching equipment loss site 2

−   isolator losses site 1

−   radome loss site 1

−   diffraction losses

−   reflection losses

Total losses site 2 *to site* 1 =

−   free space attenuation

+   antenna gain in dBi for site 2

+   antenna gain in dBi for site 1

- feeder loss site 2

- feeder loss site 1

- connector losses site 2

- connector losses site 1

- branching equipment loss site 2

- branching equipment loss site 1

- isolator loss site 2

- radome losses site 1

- radome losses site 2

- diffraction losses

- reflection losses

B.10 Rain attenuation

$$Attenuation \: / \: km = 10^{\left[1.3116 \log^{(mm \, / \, hr)} + [\: 1.5336 \log \frac{1}{F} \:] \frac{13}{F} \right]}$$
$$for \: F > 3 \: GHz$$
$$F = frequency$$

$$Attenuation \: / \: km = 0 \: dB$$
$$for \: F < 3 \: GHz$$

B.11 Availability

For F > 15 dB Fade Margin

$$100\% \left[1 - (\: a - b - 6 - 10^{-7} - f - 10^{3} - 10^{\frac{-F}{10}} \right]$$

For F < 15 dB

$$100\%\left[\frac{1-(a-b-6-10^{-7}-f-10^{3}-10^{\frac{-F}{10}}+100-I^{\left(\frac{F}{10}\right)})}{2}\right]$$

Where

F = Fade Margin
f = Frequency in GHz
a = General terrain type (from ¼ for mountainous, rough and dry, to 4 for smooth and over water)
b = General weather type (from ½ for hot and humid, to ⅛ for mountainous, and dry)

B.12 Space diversity (spreadsheet)

$$I_{sd}=\frac{1.2-10^{-3}-freq-(\Delta h_{2})^{2}-10^{\left(\frac{Fade-Difflos}{10}\right)}}{D}$$

$$New\,availability=100\%-\left[1-\left(\frac{1-Avail\,after\,\dfrac{\%}{100}\,rain\,Atten}{I_{sd}}\right)\right]$$

B.13 Frequency diversity

Do this in Wave Spreadsheet after Space Diversity

$$I_{fd}=80-\left(\frac{\Delta f}{f}\right)-10^{\left(\frac{0.1F}{f-D}\right)}$$

f = Frequency in Ghz
F = Fade margin
Δf = Frequency difference

$$New\,availability=100\%-\left[1-\left[\frac{1-old\,avail\,\dfrac{\%}{100}\,after\,rain}{I_{fd}}\right]\right]$$

B.14   Hybrid diversity

$$I_{Hd} = I_{fd} + I_{sd}$$

$$New\ availability = 100\% - \left[1 - \left[\frac{1 - avail\ \dfrac{\%}{100}\ after\ rain}{I_{Hd}} \right] \right]$$

Inde

Bibliography

Advanced Digital Systems, Kamilo Feher, Prentice Hall, 1987

The ARRL Antenna Book, ARRL, 1984

The ARRL Handbook, ARRL, 2000

Communications System, Simon Haghin, Wiley, 1986

The Complete Modem Reference, Gilbert Held, Wiley, 1991

Data Communications Testing and Troubleshooting, Gilbert Held, Van Nostrand/Reinhold, 1992

Electronic Engineers Handbook, Fink Christiansen, McGraw-Hill, 1986

Engineering Considerations for Microwave Communications Systems, GTE, 1970

Introduction to Satellite Communications, Bruce R. Elbert, Artech House 1987

The Ku-Baud Satellite Handbook, Mark Long, SAMS, 1987

Land Mobile Radio Systems, R.J. Holbecke, IEE, 1985

Mobile Communications Engineering, William C.Y. Lee, McGraw-Hill, 1982

Radio Communication Handbook, RSGB, 1999

Satellite Communications Systems, B.G. Evans, IEE, 1987

Satellite Communications Technology, Robert L. Douglas, Prentice Hall, 1988

Telecommunication Transmission Handbook, Roger Freeman, Wiley Interscience, 1981

Printed and bound by CPI Group (UK) Ltd, Croydon, CR0 4YY

03/10/2024

01040336-0017